Perfect Mechanics

Instrument Makers at the Royal Society of London in the Eighteenth Century

RICHARD SORRENSON

Docent

Press

Docent Press
Boston, Massachusetts, USA
www.docentpress.com

Docent Press publishes books in the history of mathematics and computing about interesting people and intriguing ideas. The histories are told at many levels of detail and depth that can be explored at leisure by the general reader.

Cover design by Brenda Riddell, Graphic Details.

Produced with TEX. Textbody set in Garamond with titles and captions in Bernhard Modern.

Contents

List of Figures

List of Tables

Acknowledgements

At Princeton University: Charles Gillispie who introduced me to the history of science with masterful survey courses; Mike Mahoney and Norton Wise who taught me much about the importance of the technological and the mathematical worlds; Lawrence Stone who taught me English history and was an inspirational figure; fellow students (now teachers) of English history and history of science, Ken Arnold, D. Graham Burnett, Geoff Clark, Vivien Dietz, Andrea Rusnock, and Emily Thompson. At Cambridge University: Simon Schaffer, Jim Secord, and Liba Taub for their help and warm welcomes. At the Royal Society of London: the librarians, particularly Alan Clark and Keith Moore; and the Society in general for its outstanding support of the history of science. At Oxford University: Jim Bennett who is distinguished both as a historian and a museum curator and has been very helpful over the years. At Indiana University: Kevin Grau, Elizabeth Green Musselman, John Powers, and Julianne Tuttle all of whom were a pleasure to work with and livened up many a graduate seminar. At the Dibner Institute: the Dibner family for their generous support of the history of science, Jed Buchwald for his leadership, and Jim Voekel for his gracious hosting of poker games. At the Docent Press: Scott Guthery and Mary Cronin, for their tremendous help and kindness. At Auckland: my supportive parents Keith and Judith; my delightful children Claire, Peter, and David; and my wonderful wife, Helen Sword.

The following journals are acknowledged for having previously published material that appears in this book:

- "Towards a history of the Royal Society in the eighteenth century," Richard Sorrenson, *Notes and records of the Royal Society of London,* 1996, *50*:29-46.

- "The ship as a scientific instrument," Richard Sorrenson, *Osiris,* vol. 11, 1996.

- "George Graham, visible technician," Richard Sorrenson, *British journal for the history of science,* 1999, *32*:203-221.

- "Dollond & Son's pursuit of achromaticity, 1758-1789," Richard Sorrenson, *History of science,* 2001, *34*:31-55.

Introduction

> "If anyone could work with the greatest exactness, he would be the most perfect mechanic of all." Isaac Newton, *Principia* [1]

In December of 1675, in a desperate race with Christiaan Huygens over a patent for a spring-regulated watch, Robert Hooke characterized the clockmaker Thomas Tompion as a "Slug," a "Clownish Churlish Dog," and a "Rascall" because Tompion was too slow in making a watch of Hooke's design.[2] It was Hooke's watch, not Tompion's. Robert Hooke, Fellow of the Royal Society (F.R.S.), was the patron and Thomas Tompion, shopkeeper, was the client. Fifty years later Tompion's apprentice, George Graham, similarly made watches, clocks, and quadrants for Fellows of the Royal Society; yet these instruments were known as Graham's clocks and Graham's quadrants. Language such as Hooke used towards Tompion was inconceivable towards Graham. He was not only a Fellow but also a member of the Society's governing Council and the author of several significant papers published in the Society's journal, the *Philosophical Transactions.* His testimony on, and experiments in, astronomy, magnetism, horology, and metrology were unquestioned and he was buried at Westminster Abbey with an inscription that stated his "inventions do honour... the British genius" and his "accurate performances, are the standard of mechanical skill."[3] Yet, in the early decades of the eighteenth century, one could still go to his shop in London's Strand and buy a watch or a clock from him. Like Tompion, George Graham, F.R.S., was a shopkeeper.

Nor was he alone in the eighteenth century at that supposed bastion of gentlemen, the Royal Society. Nearly two-thirds of the membership had to work for a living in one way or another; some rather grandly as high government officials, senior army officers, clerics with ample livings, and physicians and lawyers with large and successful London practices; others more modestly as sailors, surgeons, apothecaries, schoolteachers, engineers, attorneys, and instrument makers. These members tended to be active at the Society—receiving many of the Copley Medals (the Society's highest award for scientific work) and contributing extensively to the Society's journal, the *Philosophical Transactions*—and eminent instrument makers were no exception.

In the eighteenth century, the terms "instrument" and "instrument maker" were used rather than the modern terms "scientific instrument" and "scientific instrument maker." An instrument maker was usually identified as practicing one of three branches (mathematical, optical, or philosophical), although these categories blurred as the century progressed.[4] Instrument makers benefited from the expansion of science into new public spaces (coffeehouses, demonstration lecture halls, certain scientific societies) that were patronized by a fairly wide range of English society.[5] Here, the craftsmen who made the mathematical and optical instruments that undergirded the new experimental science flourished; supplying equipment, sometimes giving lectures, and very occasionally making contributions to natural philosophy themselves. At the Royal Society, artisans and other non-gentlemen such as George Graham spoke for themselves on important topics and with a broad and unfettered intelligence. Rather than languishing, like instrument makers in other times and places, as invisible technicians, they were able at times to invert the social order, resisting gentlemanly appropriation of their work and prevailing over their social superiors in scientific disputes.[6] Moreover, they ensured that the Society, far from experiencing the decline that many historians casually allude to, remained a vital and growing institution throughout the eighteenth century, carrying out work that was respected all over Europe.

The centrality of instrument makers and other men who had to work for a living, to the Royal Society, and indeed to all of British science, has been recognized by a number of scholars.[7] John Heilbron notes that "all of the productive English electricians in the 1740s came from the higher trades: Watson the apothecary, Wilson the painter, and... Ellicott [the] clockmaker." Douglas Cardwell estimates that nearly one-half of the hundred or so "leading British scientists" of the eighteenth century had to work for a living, many in quite humble positions. And Margaret Jacob argues that in eighteenth-century Britain there arose a "new social space" for engineers and their genteel employers to discuss "matters technical, applied and occasionally theoretical."[8] Yet we still know very little about many of these men's lives and working practices. Aside from brief obituaries in *The Gentleman's Magazine*, they had no Aubrey or Boswell to record their lives and prejudices; nor, with very few exceptions, did their family firms survive long enough to provide extensive records of their businesses.[9] Where they have left behind traces in the scientific institutions with which they dealt, the brief letters or bills do not tell us of their education or their private motivations. Such traces, however, along with their contributions to the *Philosophical Transactions*, do give us insight into the crucial role that these men played in making science real in the eighteenth century.

The patron saint of these men was Francis Bacon, whose aphoristic prescriptions for the practice of a sound science based on the union of thinking and acting were brought to successful fruition in the making of eighteenth-century instruments in London.[10] It was, Albert van Helden notes, "only at the beginning of the eighteenth century that instrument makers across the board were working at the very frontiers of precision technology."[11] They demonstrated a "closer and purer league" between the "experimental and the rational" faculties; thus they resembled Bacon's moderate bee, which "gathers its material from the flowers of the garden and of the field, but transforms and digests it by a power of its own," in contrast to the ant-like men of experiment who only "collect and use," or the spider-like reasoners "who make cobwebs out of their own substance."[12] Since instruments are material entities that straddle scientific and technical realms, the makers of novel or improved ones had to operate in both worlds to be successful. They needed enough science to know how and why an instrument worked and also enough technique to put designs into practice. This is not an easy business. Ants and spiders do not casually transform into bees and the interests, experience, education, proclivities, income source, and social positions of experimenters and reasoners do not often overlap.[13]

Into the class of dogged ants, Bacon placed the hidebound mechanic who, "not troubling himself with the investigation of truth, confines his attention to those things which bear upon his particular work."[14] But the eighteenth-century mechanics of whom I write in this book were very different from those Bacon describes. They did not, ant-like, pile up copies of instruments that were just like those of their fathers or the masters to whom they had been apprenticed. They set out to improve themselves and their artifacts and they did so, in part, by adopting the new experimental philosophy of the previous century. They were scientifically astute and mechanically brilliant. Like Bacon's bees, they gathered the raw pollen of their scientific and technical knowledge and transformed it into the honey of an improved science. Unlike other tradesmen, who proved unable or unwilling to provide the Royal Society with detailed—or, in some cases, with any—accounts of their various trades, instrument makers stood to gain much from advertising their skills and wares in the pages of the *Philosophical Transactions*.[15] One of them, John "Longitude" Harrison, has recently had his fame refurbished.[16] Others, while little known today, once had a national and even international eminence.

Elite instrument makers were thus similar to the painters and sculptors for whom success, according to Sir Joshua Reynolds in his 1776 address to prize-winning students at the Royal Academy, depended not only on "the industry of the *hands*, but of the *mind*. As our art is not a divine *gift*, so neither is it a mechanical *trade*. Its foundations are laid on solid science."[17] Just as artists could not rely on skill alone, instrument makers elected to the Royal Society could not make products that were *merely* mechanical, that is to say, made without thought or invention. Instrument makers too, had to have foundations "laid in solid science" and their virtues—dexterity, exactness, a respect for empirical truth—were thought to typify one particular strand of Britishness.

For example, Jesse Ramsden, who made optical and mathematical instruments and sold them in his Piccadilly shop, was awarded the Copley Medal in 1795 for his accurate design and construction of various astronomical, navigational, and surveying instruments. Sir Joseph Banks, in his presidential encomium awarding the medal, spoke of Ramsden's "Promethean touch" and his superior knowledge of optics, mechanics and mathematics. This combination of manual and intellectual skills, Banks argued, had greatly raised the "respect of foreigners for the national science of Great Britain." Ramsden himself, Banks went on to note, constructed instruments that gave "credit to the national character of Englishmen" and were accurate enough to satisfy even the "severe, but judicious scrutiny" of the Royal Society's greatest scientific rival, the Royal Academy of Sciences in Paris.[18] Ramsden was by no means unique in being so honored by the Society he belonged to. He stood at the end of a venerable tradition, spanning most of the eighteenth century, whereby elite English instrument makers were recognized as important members of the Royal Society. Their innovative instruments embodied, as did the makers themselves, the "manly" English virtues, so esteemed by the Royal Society, of being accurate, observant, mechanically ingenious, and sticking to the facts. Prominent makers were also respected for their role in improving and extending one of the central sciences in Britain in the eighteenth century—mixed mathematics—as well as for their skilful use of the Royal Society's favorite methodology—the experimental philosophy.[19]

Contemporary writers saw a promising future for English science in the eighteenth century. In his *Cyclopaedia* of 1728, Ephraim Chambers wrote a dedication to King George II that trumpeted:

> Your Majesty commands a people capable of anything. . . . We have somewhat of the boundary that circumscribes our present prospect; and separates the known from the unknown parts of the intelligible world. Under your Majesty's princely influence and encouragement, we promise ourselves this boundary will be removed, and the prospect extended far unto the other hemispheres.—Methinks I see trophies erecting to your Majesty in the yet undiscovered regions of Science; and Your Majesty's name inscribed to invention at present held impossible.

As prophecies go, this was rather a good one. As I will show in the following chapters, developments in mixed mathematics and instrument making did indeed lead into "yet undiscovered regions of Science."

The mixed mathematics that these makers both practiced and promoted was a science of central importance in the eighteenth century and yet an ancient one. As John Heilbron has argued, taking up a position outlined by Thomas Kuhn, mixed mathematics can be compared to the "classical" sciences.[20] Not only was mixed mathematics central to the Royal Society, it was the only science the British state took any interest in; directly supporting astronomy, geography, navigation, surveying and hydrography through the Royal Observatory at Greenwich, the Board of Longitude, and the Admiralty.

The importance of instrument makers to these sciences and the institutions that relied on them may still be seen today at Greenwich itself. In the last years of the nineteenth century, a new observatory building was added. The designers of the elegant brick and terracotta structure, built on the plan of a Celtic cross, looked back over the history of the Observatory and inscribed on the new building the names of the men whom they judged to have been responsible for its successes. Standing at the four points of the cross are the bulwarks of English positional astronomy, the four Astronomers Royal: Flamsteed, Bradley, Maskelyne, and Airy. In between lie the names of the instrument makers who made their observations possible including, from the eighteenth century, the telescope-makers Dollond and Herschel; the quadrant-makers Graham, Bird and Ramsden; and the chronometer and clockmakers Harrison, Arnold and Earnshaw. These makers are memorialized for more than being clever with their hands. They reached their elevated status by thinking, mathematizing, experimenting and, within the bounds of good taste, promoting their inventiveness.

The science of the eighteenth century, grateful heir to the new experimental and mechanical philosophy of the preceding century, was self-consciously and proudly dependent on a variety of instruments that extended the range and power of the unaided human senses.[21] Some—the telescope, microscope, thermometer, barometer, pendulum clock, electrical charge generator, and air pump—

were modern; others—the ruler, balance, and compass—were ancient. But they were all material objects and they all had to be made by hand. In a rather complex way, then, the new science was articulated by craft skill.[22] A dependence on instruments implied a dependence on instrument makers, who were usually artisans. And this could be an awkward thing that left many troubling questions for the typical practitioner of science at an eighteenth-century academy, university, or court, who was usually genteel, university educated, and commercially disinterested. Did the lower social class, lesser educational achievement, and greater commercial self-interest of the makers necessarily disqualify them as knowledge producers? Would the instrument work as advertised? Did the quality of an observation depend, in part, upon the instrument itself? Did it matter who made a particular instrument?

Dependency rarely fosters gratitude, and the few written traces that instrument makers have left tend to be in the letters of professors or academicians or genteel natural philosophers who express anger that an instrument they want has not yet arrived, or nervousness that it may never materialize. Occasionally, the writer speaks of the maker with genuine affection and esteem. More often, the tone is one of reluctant flattery alternating with exasperation at being reliant upon a tradesman to be successful in the business of natural philosophy. But get these instruments you must if you wished to demonstrate that you were a serious observer of nature in the eighteenth century. If you were ambitious for recognition but lived on the periphery of Europe, you had no choice but to buy the artifacts, ready-made, from the centers of production in London, or Paris, or Amsterdam; the local tradesmen would not do.

For those who lived more centrally, perhaps a skilled maker of instruments could be visited and ordered to execute a design. But would he understand your instructions? Could you, in turn, understand what he was actually capable of making? Who deserved the credit for a successful and innovative instrument or, for that matter, for the resultant knowledge its use created? In some cases, usually when the instrument was well understood and widely available, full payment for the goods ended the transaction between maker and user. The former was happy to have identified and satisfied a market for his products; the latter left the shop in possession of an instrument he intended to use just as he pleased. In other cases, however, usually when the instrument was novel or difficult to make, a sale did not end the obligations between the buyer and seller. Just as seventeenth-century Lucasian professors or Medician courtiers or Parisian Academicians sought more than a salary as compensation for their work, so too many eminent eighteenth-century instrument makers wanted more than merely to have their bills paid on time. They also wanted credit for their part in adding to the stock of knowledge about the natural and the artificial world.[23]

This book examines the crucial role played by eighteenth-century scientific instrument makers in advancing English science. Chapter One offers a general overview of the instrument-making trade in eighteenth-century London. Chapter Two goes on to provide a detailed analysis of the social and intellectual structure of the Royal Society, thereby arguing that the neglect of the Society's history between the famous, long, and well-chronicled presidencies of Sir Isaac Newton (1703-1727) and Sir Joseph Banks (1778-1820) is not justified. There are no book-length studies of the eighteenth-century Society and very few scholarly papers, quite in contrast to the amount of literature on the Society in the surrounding seventeenth and nineteenth centuries. The chapter will also make clear why certain instrument makers were allowed to be Fellows in the eighteenth century and why they were so important to its functioning. Their instruments were crucial to two of the Society's main activities—measurement and experiment—for by the early eighteenth century it was a "commonplace that instruments had an essential place in the study of nature."[24]

As Chapter Three will show, London instrument makers were highly valued because they supplied accurate and innovative instruments capable of making careful measurements all over the globe. The increasing accuracy of their instruments had significant philosophical as well as pragmatic implications. Firstly, since mathematical instruments "worked with quantifiable entities—distance, angle, and time," which were "literally incorporated into the instruments that measured them," they evermore closely approached the ideal entities of space and time which, along with mass, were the underlying basis of the mechanical philosophy.[25] Secondly, a more precise instrument could reveal phenomena that were completely unexpected; thus making new contributions to natural philosophy.

But instrument makers did not only make to measure; they also measured to make. They were, as I argue in Chapter Four, dedicated to a rational instrument design. Only by deliberate and systematic measurement of the properties of the materials from which the instruments were constructed could new ones be made that actually approached the geometrical ideal their makers strived for. The shape of the instrument was only the beginning; it also mattered what material it was made from. A lens could be shaped as closely as possible to the geometric and optic ideal but, if the qualities of the glass were not well understood, the lens might not bend light as theory demanded. A pendulum could be constructed to a specific length to beat out as accurately as possible a second of time on one day, yet the same pendulum would then beat out a different time on a colder (or warmer) day. Once the variant nature of matter was measured and understood, geometrical principles could then be used to design and make instruments that were themselves invariant: clocks that beat time uniformly over temperature ranges or telescopes that focused images without creating distracting colored fringes.

Instrument makers in eighteenth-century London walked a fine line between gentility and commerce. But the precarious middle ground they inhabited all but disappeared with the French Revolution. Those with sufficient wealth and insufficient ambition retired from the fray to enjoy life as independent gentlemen.[26] Those who had to continue working for a living suffered a corresponding decline in scientific status.[27] In their attention to business, as we shall see in Chapter Five, instrument makers no longer had the time or education to make polite contributions to science at the Royal Society; now, their function was a clearly subordinate one of supplying instruments for others. The last of the eminent eighteenth-century makers, Jesse Ramsden, was honored by the Society for his astonishingly accurate mathematical instruments, but he had no control over their use, nor did he lay claim to any new discoveries in natural philosophy. Even those instrument makers who had made such discoveries, moreover, suffered a substantial decline in scientific reputation by the end of the century. John Dollond, who had been hailed in 1758 as the inventor of the achromatic lens and the discoverer of a new and major principle in optics demonstrated by extensive experimentation, had his claims to both contributions deliberately demolished, while George Graham's crucial roles in discovering the aberration of starlight and measuring the shape of the earth were all but forgotten.[28] In explaining why these instrument makers were so revered in their own day, this book seeks also to resurrect their well-deserved fame.

Chapter 1

Markets and Makers

1.1 Markets

The London trade in instruments throughout the eighteenth century was the largest of any city in Europe.[29] On the supply side, driving this trade, were a domestic brass and glass industry supplying raw materials that were at least as good and cheap as any in Europe; a flexible and well-capitalized system of sub-contracting; a loosely controlled apprenticeship system that did not restrict workshop size; a vigorous and innovative retailing trade supplying the huge London market; an easy availability of popularized natural philosophy, which instrument makers understood and made use of; a mechanically skilled and ambitious provincial labor pool; and a craft tradition in mathematical instrument making and watchmaking that stretched back to the late sixteenth century.[30] On the demand side were the British state buying gauging instruments for the customs and excise, marine instruments for the navy, and astronomical instruments for the Royal Greenwich Observatory; domestic, colonial, and European consumers purchasing marine, surveying and household instruments; natural philosophers in Britain and abroad ordering experimental and observational instruments; and popular lecturers and schoolteachers buying a whole range of demonstration instruments to explicate the new natural philosophy.[31]

The practitioners of this philosophy used mathematical instruments in novel ways and invented two entirely new types of instruments—optical and philosophical—for carrying out experiments.[32] Indeed, this outburst of inventive activity provides one of the most compelling reasons to talk about a "scientific revolution" at all. As Derek Price has noted, "the scientific revolution, as we call it, was largely the improvement and invention and use of a series of instruments of revelation that expanded the reach of science in innumerable directions."[33] These instruments (among them telescopes, microscopes, prisms, barometers, spring- and pendulum-regulated timekeepers, thermometers, and air-pumps) were of crucial importance to the new philosophy of the seventeenth century, as Robert Hooke made clear in 1665 in the Preface to his *Micrographia*:

> ...the next care to be taken in respect of the Senses is a supplying of their infirmities with *Instruments*, and, as it were, the adding of *artificial Organs* to the *natural* ...from whence there may arise many admirable advantages towards the increase of the *Operative* and the *Mechanick* Knowledge. ...The truth is, the Science of Nature has been already too long made only a work of the *Brain* and the *Fancy*; It is now high time that it should return to the plainess and soundness of *Observations* on *material* and *obvious* things.

By the eighteenth century, the new instruments listed above had become uncontroversial, replicable, and readily available commercially. Instruments were used to make facts and theories and, in turn, themselves became what Derek Price calls "embodied explanations of the way things worked."[34]

The flourishing London trade in mathematical instruments was soon organized under a guild structure, although since the guilds were weakening by the early seventeenth century, the makers were not forced to belong to a specific guild, nor did they seem to take much notice of regulations that governed the number of apprentices each guild member was allowed in his workshop.[35] Most mathematical instrument makers joined the Clockmakers' Company, founded in 1631, or the much older Grocers' Company which was chartered in 1345.[36] The new instruments of natural philosophy were produced by optical and philosophical instrument makers. The former tended to belong to the Spectaclemakers' Company after its founding in 1629, the latter to no particular guild at all.[37] These two kinds of makers were separated not by the materials they worked with—both used glass extensively—but more by their techniques. While optical instrument makers worked with ground glass to make lenses for telescopes and microscopes, philosophical instrument makers worked with blown glass to make air pumps and barometers.

In the eighteenth century these divisions blurred as mathematical instrument makers fitted their quadrants, sextants and sectors with telescopic sights. By mid-century R. Campbell, in his *London Tradesman*, included no separate division for the philosophical instrument maker, noting that the occupations of mathematical and philosophical makers could no longer be easily distinguished:

> ... [the] Mathematical-Instrument-Maker makes all kind of Instruments constructed upon Mathematical Principles, and used in Philosophical Experiments: He makes Globes, Orrerys, Scales, Quadrants, Sectors, Sun-Dials of all Sorts and Dimensions, Air-Pumps, and the whole Apparatus belonging to Experimental Philosophy. He ought to have a Mathematically turned Head, and be acquainted with the Theory and Principles upon which

> his several Instruments are constructed, as well as the practical Use of them. . . . It is a very ingenious and profitable Business, and employs but few Hands as Masters. The Journeymen earn a Guinea a Week, and some more, according as they are accurate in their Trade.[38]

At the beginning of the eighteenth century, some vestiges of the guild structure were still in place; for instance, although instrument makers designed and manufactured the subcontracted parts with a few apprentices and assistants, the maker would generally finish the instruments himself and put his name to them. By mid-century, the maker's role was largely a supervisory one, as attested by Campbell's 1747 description of a watchmaker's shop:

> . . . he scarce makes any thing belonging to a Watch; he only employs the different Tradesmen among whom the Art is divided, and puts the several Pieces of the Movement together, and adjusts and finishes it. . . . Engines [are used] for cutting the Teeth in the several Parts of the Movement, which were formerly cut by Hand. . . . It is supposed, however, that he can make all the Movements. . . . He must be a Judge of the Goodness of the Work at first Sight, and put his Name to nothing but what will stand the severest Trial; for the Price of a Watch depends upon the Reputation of the Maker only.[39]

By the end of the century, instrument-makers' workshops had grown even larger and the signature of the maker on an instrument had become a trademark. The quality of the instrument was now guaranteed by the reputation of the maker rather than by his actual work on the instrument.

In some cases, the changes in working practices were even more profound. At the beginning of the century, mathematical instrument makers divided their instruments themselves, ensuring the accuracy of the assembled instruments by geometrical methods. By the end of the century, however, the accuracy of some of these instruments was ensured by machine tools. Allan Chapman gives a wonderfully detailed study of the techniques of instrument division and concludes that the "division of the quadrant was a geometrical exercise, but the division of the full-circle became a piece of mechanical engineering." Mathematical instrument makers of the early and mid-eighteenth century "worked within a tradition that may be termed classical, whereas Ramsden and Edward Troughton are best identified with that tradition which was to produce the Industrial Revolution."[40] This progression contributed to the decline of the scientific reputation of instrument makers within the Royal Society, as they became more involved in the problems of machine tool-building rather than natural philosophy. Only three makers belonged in the nineteenth century and they all had apprenticeship or family links to the great makers of the eighteenth century.[41] Beyond the Royal Society however, the techniques of accuracy that these makers established went on to be used by machine toolmakers in the developing Industrial Revolution of the following century.

The growth in the instrument trade and the flexibility of the guild system are indicated by how few eminent eighteenth-century instrument makers were born in London and learned instrument making as their first trade there.[42] A rather remarkable number of them began their working lives in other trades. For example, John Bird and Jesse Ramsden began as cloth workers in the north of England before coming to London to serve apprenticeships under other instrument makers. John Dollond was also a cloth worker, in Spitalfields, London, before he joined his son in an optician's business. John Harrison was a Lincolnshire carpenter and self-taught clockmaker who came to London to chase the Longitude prize of £20,000. John Smeaton moved to London from Yorkshire to become an attorney but switched to instrument making before moving on to a highly important career in civil engineering.

That guild regulation should limit such men to practicing only the trades in which they were formally trained, was a doctrine which found little sympathy with English judges when guilds sought to defend their privileges. For example, Lord Camden (Charles Pratt, F.R.S. 1742), deciding in 1763 in the favor of a defendant who had been sued by a guild for following more than one trade, used the illustrious example of John Harrison in making his ruling:

> *Mr Harrison of Red-Lyon-Square* served an apprenticeship to the trade of a *carpenter*, but for twenty six years he has been a watch-maker, and though he never served as an apprentice to the trade... is the best maker of timepieces in the world... and shall this man be hindered from making watches and exercising the trade of carpenter also if he pleases?[43]

Of those makers who were trained in instrument making but grew up in places other than London, George Graham came from Cumberland to be an apprentice to a London watchmaker, while Edward Nairne came from Kent to be apprenticed to a London optician. James Short and John Whitehurst were already skilled makers before they came from Scotland and Derby, respectively, to find fame and fortune in London. William Herschel, who grew up in Germany and settled in Bristol, was unusual in that he did not even have a London shop. Like John Dollond, Herschel was self-taught in optics; following his 1781 discovery of Uranus, he made a substantial living from the sale of his reflecting telescopes.

The developments in the instrument-making trade outlined above led to a market in eighteenth-century London that may be usefully analyzed into three broad areas (Tables 1, 2, and 3). The market for metrological instruments was the only one to exist before the scientific revolution. The "pre-scientific-revolution" instruments in this class included magnetic compasses, quadrants, globes, levels, drawing instruments, gauges, and balances. Pendulum clocks, telescopes, barometers and thermometers were added in the seventeenth century. The most important newcomers in the eighteenth century were chronometers, dividing engines, large theodolites, and achromatic lenses, all of which became commonly available after they had first been developed in response to the demands of the British state. A few elite makers supplied the "special" part of this market, primarily in response to orders and prizes from the state.[44]

Metrological/General	chronometers, compasses, globes, sextants, quadrants, theodolites, levels, drawing instruments, gauges, balances, hygrometers, weights, rules
Metrological/Special	chronometers and dividing engines for the Board of Longitude; pendulum clocks, quadrants and sectors for the Royal Greenwich Observatory; theodolites for the Ordnance Survey; standard yard for the Royal Society
Natural Philosophical/Investigative	pendulum clocks, globes, sectors, quadrants, micrometers, pyrometers
Natural Philosophical/Lecturing	orreries, globes
Household	clocks, orreries, globes

Table 1: Markets for Mathematical Instruments

The market for instruments that investigative natural philosophers used came into existence in the seventeenth century and it contained newly invented instruments (pendulum clocks, telescopes, air pumps) as well as older mathematical and optical instruments used in the new context of natural philosophy (quadrants, magnetic compasses, prisms). New instruments were added in the

Metrological/General	telescopes
Metrological/Special	telescopes and achromatic refracting lenses for the Royal Greenwich Observatory
Natural Philosophical/Investigative	telescopes, microscopes, optics for mathematical instruments, burning lenses
Natural Philosophical/Lecturing	burning lenses, prisms
Household	spectacles, telescopes, microscopes, prisms, magic lanterns, opera glasses, camera obscura

Table 2: Markets for Optical Instruments

eighteenth century; most notably electrical machines and the apparatuses of pneumatic chemistry. The market for instruments that popularized natural philosophy was newly created in the late seventeenth and early eighteenth centuries, as natural philosophy moved out of the courts and academies of princes and into the newly burgeoning marketplace of consumers. Household instruments exemplified the new philosophy: an orrery, the clockwork universe; a pendulum, the regular force of gravity; a barometer, a history of the air.[45]

Instrument makers who did nothing to advance natural philosophy or precision mixed mathematics, while they could flourish supplying various parts of the instrument market, were not considered for membership of the Society. Makers were still not considered for election unless they had also made new, improved or highly accurate instruments; undertaken experiments to improve their instruments; or published papers in the *Philosophical Transactions* describing experiments and observations carried out with their instruments. These qualifications were not explicitly laid out by the Royal Society; they result from my analysis of those instrument makers who were elected as Fellows. Examples of large

Metrological/General	barometers, thermometers
Metrological/Special	telescopes and achromatic refracting lenses for the Royal Greenwich Observatory
Natural Philosophical/Investigative	air pumps, pneumatic and chemical apparatuses, magnets, compasses, electrical machines
Natural Philosophical/Lecturing	air pumps, pneumatic and chemical apparatuses, magnets, electrical machines, apparatuses to demonstrate principles of mechanics, models of machines, mechanical "toys"
Household	thermometers, barometers, air-pumps, electrical machines, mechanical "toys"

Table 3: Markets for Philosophical Instruments

and successful family firms who supplied all parts of the market in varying degrees but did not satisfy the above conditions and whose individual members did not apply for membership include the Sissons, the Adamses and the Joneses.[46]

The character of the market, and of instrument making itself, changed radically in the nineteenth century. Gloria Clifton and Eva Taylor, in their books about British instrument makers, could barely keep track of the burgeoning number of makers in the first half of the nineteenth century and not at all in the second half.[47] As makers became retailers or manufacturers to cope with the increased demand and competition, their scientific status, as I shall argue at the end of this book, disappeared.[48] A relative decline in the international superiority of English instrument making by the mid-nineteenth century was caused by a mixture of complacency, attacks from overseas competitors, pressures to industrialize output, and reduction in the scientific knowledge and status of the makers themselves. Eighteenth-century instrument makers were situated in a "peculiarly sensitive position, poised between the changing patterns of scientific and industrial development." English makers seem to have lost their poise to the extent that, eventually, "full membership of the scientific community... became impossible."[49]

1.2 Makers

Full membership of the scientific community had been equally unattainable for instrument makers in the late seventeenth century. But because of the existence of Robert Hooke's diary, we can get some insight into how makers such as the London watch and clockmaker Thomas Tompion began to improve themselves sufficiently (learning how to make novel mathematical instruments and operate within a scientific culture) such that their successors in the next century could attain a more exalted status.[50] Hooke mentions meeting with Tompion at least 300 times from 1674 to 1693, and during that time Tompion constructed many of Hooke's contrivances, including, most importantly, his quadrant and his spring watch.[51] Since we have only Hooke's account of their meetings and not Tompion's, it is at times difficult to interpret their relationship. One thing, however, is fairly clear. The relationship, as far as Hooke was concerned, was not one between social or intellectual equals. Hooke certainly introduced Tompion to that part of the scientific world of London that was to be found in various coffeehouses: Blacklock's, Garaway's, and Man's among others. His diary is scattered with entries like the following for Sunday, July 5, 1674: "Received home quadrant from Tompion. Sir Jonas More and Tompion here and at Blacklock's"; or that for Friday, July 9, 1675, which stated: "At Garaway's with Whistler, Flamstead, Tompion, Hill etc."[52] Hooke also lent scientific books to Tompion, possibly teaching him about the new natural philosophy that the Royal Society espoused. On June 16, 1675, Hooke lent Tompion "Fosters *Miscellanys* and Streets *Astronomy*."[53] The first book was about simple astronomical instruments and the art of dialling, but the second was a more theoretical astronomical text. Hooke also undertook lengthy "discourses" with Tompion about the mechanics of clocks and watches; in these discussions Hooke, though reluctant to acknowledge it in his diary, must have been more the pupil than the teacher. Further, he often acted as a middleman selling Tompion's watches to other virtuosi. He acquired a very expensive example for Sir Christopher Wren, beating Tompion down to a final price of £8 "for the watch with seconds."[54]

Why did Hooke spend so much time with, and effort on, this London clockmaker? In the first place, Tompion seems to have been a diplomatic listener, patiently allowing Hooke to lecture him about his own business: "With Tompion at Man's. Told him about my way of springs by a hammer and anvil. . . which he approved of. I conceive it the best in the world."[55] Secondly, Hooke learned much from Tompion about the properties of materials and the intricacies of clockwork as when, on Sunday October 4, 1674, the two men spent all day discoursing "much. . . of the strength of the Spring. . . Of hammer hardening steel or iron. . . that Iron is harder to melt than either gold or copper. . . about grinding glasses."[56] Thirdly and most importantly, Tompion was able to make Hooke's "inventions" real. To get them made, and made in time, Hooke had to teach Tompion the reasons behind the design, and keep him hard at work. Eventually, however, following Hooke's frantic efforts to forestall Huygens's claim to the invention of the spring-regulated watch, their relationship deteriorated. Hooke, with his usual paranoia, suspected Tompion of trying to steal his ideas and of no longer faithfully working for him alone; thus on Sunday April 9, 1676, he reports that Tompion was trying to extract enough knowledge from him to make watches on his own: "Tompion said he would ingage etc. but was but to pump".[57]

Despite their disagreements, Hooke inspired Tompion to make a quadrant and a spring-regulated watch, knowledge that was later acquired and improved upon by Tompion's apprentice, George Graham. A watchmaker was the obvious tradesman to turn to for a horological invention, but why did Hooke use Tompion and not an established mathematical instrument maker who would have specialized in making quadrants? The choice is more logical than it might at first appear. Watchmakers, in order to make gears, were skilled in practical geometry. They were capable of fine and accurate work with brass and steel. Since precision clockmakers had to set their clocks against the stars, they had a working knowledge of dialling and astronomical theory. Hooke was not disappointed in Tompion, who proved to be a quick worker and learner. In his pamphlet announcing his new quadrant, Hooke recommends that:

> ... any person [who] desires one of [the quadrants] to be made, without troubling himself to direct and oversee a Workman... may imploy *Mr. Tompion*, a Watchmaker in *Water-Lane*... [who] hath seen and experienced the Difficulties that do occur therein, and... [is] very careful and curious to observe and follow Directions, and to compleat and perfect his Work, so as to make it accurate and fit for use.[58]

Hooke's quadrant is historically important not only because it marked the beginning of Tompion's career as a maker of instruments for the virtuosi at the Royal Society, but also because the instrument itself was designed to unprecedented levels of accuracy through the use of telescopic sights and micrometer screws. It also became the object of a violent dispute between Hooke and John Flamsteed, the first Astronomer Royal at the Greenwich Observatory. Flamsteed thought enough of the instrument to use it for over a year at the Observatory. But when Hooke maliciously purloined it, the enraged Flamsteed deprecated the designer, Hooke, and therefore the design since "I lost nothing by it; for it was so ill contrived by him, that I could not make it perform.[59]" However, Flamsteed couldn't carry on his business without instruments and it was the more compliant Tompion and not the "ill-natured" Mr Hooke who went on to further supply the Observatory, making two clocks and parts of a sextant. Similarly, Tompion's successor, his apprentice and nephew by marriage, George Graham would make three clocks and a quadrant for Flamsteed's successor, Edmond Halley.[60]

But for Graham, a major change occurred in his status. The quadrant designed by Hooke, even though Tompion made it, was known as "Hooke's quadrant." The quadrant made for Halley was known as "Graham's quadrant" because he designed and oversaw the construction of the instrument. Graham's Greenwich quadrant finally realized, in 1725, Hooke's dream of half a century earlier: it had an accuracy of "minutes, seconds," if not quite of "single thirds."[61] But whereas Hooke was a middleman for the virtuosi of the seventeenth century, designing and procuring their new instruments, eighteenth-century instrument makers such as Graham knew enough natural philosophy to design and make improved instruments directly for those Fellows of the Royal Society who required them. The most knowledgeable instrument makers acquired enough science—especially a facility in mixed mathematics and experimental philosophy—to be worthy members of the Royal Society themselves.

As we shall see in the next chapter, the Royal Society was never an institution with monolithic or severely restricted intellectual aims; indeed, by attempting to make itself the arbiter of the pursuit of natural *and* useful knowledge, the Society laid claim to an extensive but fundamentally divided territory, and the strains of patrolling this uneven terrain occasionally threatened the Society's very being. In one important respect, however, the Royal Society was very homogenous. Fellows, no matter what their social origin or position which were quite varied, had to behave in a gentlemanly way. Beyond the necessary display of good manners, this requirement imposed a very important condition: to gain scientific credit, Fellows had to freely give away the fruits of their labors.

They showered each other, their peers abroad, and their Society with gifts of specimens, demonstrations, engravings, accounts, observations, models, instruments, theorems, journals, manuscripts, books, letters, papers, their administrative services, and of course, much talk. Only one Fellow, the Curator, received cash payment in return for such effort; his was always an awkward position, and the role was finally abolished in 1744. When the gift-giving was very elaborate, or the action burdensome upon the resources of the particular Fellow, the Society might lighten, but not entirely remove, the load. Thus, on the 27th of April, 1730, the Society extended to George Graham a courtesy it had extended to Isaac Newton in the previous century.[62] With a unanimous vote, the other Fellows relieved Graham of the need to pay the Society's dues and refunded his bond "in consideration of his Services to the Society."[63] To seal the mutual gift-giving, Graham delivered a clock of his own making for the Society's Meeting Room, donating "the cost of his own workmanship."[64]

For a man of independent wealth, liberality came easily; indeed, it was a mark of his gentility. But such disinterestedness caused a fundamental problem: if a man did not depend upon the knowledge he gave the Society for his livelihood, did he really know what he was talking about? Conversely, if a Fellow did have to make his living from science, just how much could he afford to give away? Was he always telling the whole truth, or just enough to advance his own economic interests while protecting his trade secrets, whether he had something to sell (instruments or a new drug) or services to offer (a surgical technique or mathematical instruction)? These questions were never entirely put to rest. They remained points of tension well into the nineteenth century, often expressed by subsequent historians as a struggle between "amateurs" and "professionals."[65]

Certainly, in the eighteenth century, the Society had its share both of inexpert gentlemen and of non-gentleman experts.[66] The non-genteel experts were listened to, and at times lauded, at the Society, not because of their social prominence, which was often negligible, but because of their professional and craft capabilities and their devotion to science. They were by no means unlettered oafs. To make their way in the brutally competitive world of Georgian London, they had to please clients, customers, and patrons who were often gentlemen or even aristocrats. Thus they adopted a reasonable facsimile of the manners, if not the mores (which they could not afford), of their social betters.[67] Self-made men, they molded themselves into a form that was founded upon their working identities (which gave them the authority of experienced knowledge) but glazed over with a sheen of politeness (which made them fit conversationalists and believable testimonialists). Sometimes the glaze cracked under the severe heat of controversy over who deserved credit for valuable scientific advances. But it took the inferno of the French Revolution and the resulting conservative reaction in Britain, along with a relentless move toward specialization in science, to shatter this unusual eighteenth-century form.[68]

However, such a change in the affairs of the Society was not apparent in the immediate decades before the French Revolution. In the middle of the eighteenth century, eminent apologists such as David Hume argued for the beneficial effects of commerce. The continued refinements of the mechanical arts, Hume asserted,

> ...commonly produce some refinement in the liberal [arts]. ...We cannot reasonably expect, that a piece of

> woollen cloth will be wrought to perfection in a nation, which is ignorant of astronomy, or where ethics are neglectcd. The spirit of the age affects all the arts.[69]

It is important to note that for Hume, as for other essayists who wrote approvingly of their modern and commercial age, the pursuit of one's own commercial interests was in no way inimical to the production of liberal, philosophical, knowledge; indeed, quite the reverse was the case. Improvements in the arts, too, would aid in the production of an admirable sociability, since the more the arts are refined and improved, the more men "enriched with science" congregate in cities "to receive and communicate knowledge."

The venues for this behavior were the "particular clubs and societies" that such men joined. In these places they not only learned of improvements that interested them (either for pleasure or business) but, through social intercourse with their fellows, became more polite. Thus, for Hume, it was in the most advanced nations, of which England was in the van, that "*industry*, *knowledge*, and *humanity* are linked together by an indissoluble chain."[70] Nor was this a chain whose links were forged in private life only. The reason of the people in general was strengthened and refined by constant application to the "more vulgar arts."[71] Such a right thinking public would then in turn construct a well-made government. Hume gives us here some quite substantial philosophical space to make room for shopkeeper instrument makers at the Royal Society of London. These men constantly refined their instruments with a careful attention to precise construction and measurement. They flocked to London to improve themselves and establish national and at times international reputations, they were sociable, and they contributed to the public good because their instruments served the British state's interests in navigation, mapping, assessing taxes, and monitoring the quality of the coinage.[72] They were part of an ascending and virtuous spiral of refinement whereby their instruments improved the sciences and the sciences improved their instruments. But Hume's analysis can be applied still further to the instruments themselves, whose most subtle refinement often lies deep within the mechanism. Many instruments depend upon the exquisitely delicate operation of balances and restraints.[73] Thus in a barometer the mercury is exactly balanced by the weight of the unseen atmosphere; in a watch the balance spring regulates the clockwork; and in a weighing device a lever balances standard weights against the object being weighed. The economy, balance, and restraint of the instruments themselves thus embodied the "regulated protocols" that marked, for Hume, "a polite and rational social identity."[74] It is to that polite and rational London club, the Royal Society, to which I now turn.

Chapter 2

The Royal Society of London

Today the Royal Society stands at the geographical heart of the British state. From the beautiful rococo library, one looks out over the Mall. To the left, the Admiralty Building and Churchill's wartime bunker back onto Whitehall and skirt the Horse Guards Parade. From here the guards and their splendid mounts troop out along the Mall to protect the Queen or more specifically her realm's tourist revenue. To the right lies Buckingham Palace, home of the Society's patron. Straight ahead, across St. James's Park, is Parliament. Behind the library lie Pall Mall and genteel clubs; nowhere is a shop to be seen. The Regency-style buildings themselves are elegant and spacious. They stand on Crown land and were built by John Nash from 1829 to 1832. Now they are let by the state to the Royal Society for a nominal sum.

No experiments take place here. Fellows come to meet other Fellows, to use the library, and to sit on committees. They and their administrative officers oversee the Society's publications and the disbursement of the Society's and the government's money to various scientists. The library and the archival record of the Society's existence are superb, mainly as a result of the Society's continuous interest in compiling histories of itself and of its members. Unable to keep up with the increasing torrent of scientific publications, today the library collects only the Society's own publications, the publications of other national scientific societies, and works about the history of science.

With each shift of residence from its initial location in the City of London, the Royal Society has moved farther away from the sight of shops and counting houses, towards the buildings that house the Sovereign, Parliament, the Departments of State, and the gentlemen's clubs. Some things, to be sure, have not changed. The Royal Society remains far smaller than the sum of its parts; the Society is famous for what its members do elsewhere, not for what they do at the Royal Society. Virtually every distinguished scientist in England is a member. The Society's constitutional structure has been little altered since its first Royal Charter in 1662. Now, as then, it is an exclusive London club validating the work of its members, keeping careful record of its own activities, and publishing the *Philosophical Transactions*. Over the years, however, the Society has performed fewer and fewer experiments and narrowed its membership criteria, even while receiving ever more of the state support it always wanted and being consulted more often by government departments. In fact, it has taken the Society three centuries to attain its present position, one that it had hoped to achieve at its foundation in 1660.

Despite the Royal Society's constant trajectory of growth and achievement, a long-standing impression persists among scholars that the Society was in decline during the eighteenth century. This perception relies on four sources: the often-stated belief that the Society failed to follow the illustrious example of its greatest Fellow, Sir Isaac Newton; the negative opinions of several literary lions of the late seventeenth and early eighteenth centuries; the continued popularity of Charles Babbage's *Reflections on the Decline of Sci-*

ence in England; and the intensive study devoted to the Society's early years which has overshadowed later periods.

Yet the declinist argument may be rebutted. Firstly, the Royal Society did in fact practice Newtonian science; but it was that kind of empiricism found in Newton's *Opticks* rather than the mathematical natural philosophy of his *Principia*. Secondly, continued reliance on a few oft-quoted opinions of literary figures seems a poor substitute for checking the satirists' own rhetorical agendas or what was actually happening at the Royal Society. For example, George Rousseau's survey of science in eighteenth-century England is typical in relying on the usual critics—Steele, Swift and Hill—for its conclusion that "the society from 1680 to 1780 fared ingloriously, except during Newton's presidency." Thirdly, Babbage's work has too often been taken at face value as an analysis rather than as the polemic it was. His criticism is based largely on his assertion that "mathematical sciences" were virtually never in evidence at the Royal Society. In England,

> ...particularly with respect to the more difficult and abstract sciences, we are much below other nations. ...To trace the gradual decline of mathematical, and with it of the highest departments of physical science, from the days of Newton to the present, must be left to the historian.

Babbage laid the prime responsibility for the "decline" at the foot of the Royal Society, which he accused of "mismanagement," and whose Fellows displayed "foibles and follies," at every turn.[75] On the contrary, as I will show throughout this book, there were many expert practitioners of mathematics at the Royal Society in the eighteenth century—but not of the sort of mathematics Babbage had in mind. Pure mathematics was almost entirely absent, yet mixed mathematics was practiced quite consistently.

And finally, the scholarly work on the early years of the Society, impressive both in extent and sophistication, should serve as an encouragement to look more carefully at later periods. Rather extraordinarily, Charles Weld's *A History of the Royal Society*, from the mid-nineteenth century, remains the best book-length study to cover the eighteenth-century Society; a period that Sir Henry Lyons deals with only briefly, and one senses reluctantly, in *The Royal Society: 1660-1940*.[76]

In fact, far from having "fared ingloriously," the Society experienced a period of exceptional productivity and growth throughout the eighteenth century, partly due to the elite mathematical or optical instrument makers who were judged by their contemporaries to be indispensable to the progress of science. In a presidential address at the end of the eighteenth century, Sir Joseph Banks called them "a body of men, on whom the improvement of every branch of natural knowledge is immediately dependent."[77] But while the Royal Society could not function properly without the mechanical productions of these men, did it have to admit them to its ranks? Instrument makers were not landed gentlemen, members of ancient professions, or scholars; they were shopkeepers, guild members, and autodidacts. Their prime interest was a commercial one—to sell their wares—so when they described an instrument, was their text more puffery than philosophy? To finesse this problem, the Society only admitted makers to its ranks who demonstrated a high moral character, a devotion to natural knowledge at least as keen as their devotion to business, and a deep knowledge of experimental methods. Furthermore, they had to be productive, observant, useful, reliable, accurate, expert, and sociable. Instrument makers whose work commanded "the respect of Foreigners," Banks noted, were among the Royal Society's most eminent members. Living an ideal scientific life and contributing to a major branch of the tree of knowledge, they filled a new social niche at the eighteenth-century Society.

2.1 Modes of Scientific Life

To the eighteenth-century Fellows of the Royal Society, the ideal scientific life was exemplified by those members who made careful observations of natural or artificial phenomena, gave them a mechanical explanation or demonstration where possible, avoided grand theory, and above all produced reliable and accurate facts with accurate instruments where appropriate. Not all Fellows could live up to this model. Seeing themselves as patrons and occasional observers rather than as regular practitioners of science, many Fellows meant to do no more than pay their dues, read the *Philosophical Transactions*, and occasionally attend meetings or correspond with the Society. In return they expected informative, entertaining, enlightening and useful knowledge. Such members were called by one President of the Society "men of wealth and station, disposed to promote, adorn, and patronize science."[78] To criticize them for not being productive thus makes as much sense as criticizing an opera audience for not singing. Instrument makers, to continue the metaphor, performed rather than patronized; they sang or, at the very least, made and designed the sets. In a Society dependent for survival upon the modest patronage of hundreds of members rather than the grand patronage of a single prince or monarch, such Fellows had their role to play. Without their £2.12s a year (or £27.6s life membership) the Society would have disappeared.[79]

The more energetic Fellows, those who "successfully cultivated science," lived an observant life with great intensity.[80] Exemplary was James Bradley, the future Astronomer Royal, who made nearly 20 years of continuous astronomical observations that established the nutation of the moon's axis.[81] More typical, but scarcely less immoderate in daily intensity, were Fellows like George Graham, who reported the results of his thrice-daily, year-long observations tracking the behavior of the magnetic dipping needle, even while he was also in the middle of a three-year daily investigation of his new heat-invariant pendulum.[82] Nor was such a lifestyle unique to the Society's Fellowship; it spread to others who, while perhaps not suited to or desiring membership, nonetheless reported or bequeathed to the Society their observations. Thus Captain Middleton presented thirty years of weather observations, the posthumous contribution of "Mr. Patrick weather glass maker in the Old Bailey." Seven years later, the mathematical instrument-maker John Bird demonstrated his devotion to science by living the observant life while still in bed. Despite being plucked from his sleep by the earthquake of March 8, 1750, he was still able to tell the Fellows that his bed was "actually lifted from the floor... twice to the right and [twice] to the left."[83]

The observant life had moral implications that the Presidents of the Society in the middle part of the eighteenth century (Martin Folkes and Lord Macclesfield) made clear during their annual Copley Medal speeches. This prize was, and still is, the Society's highest honor. A bequest from Sir Godfrey Copley at the turn of the eighteenth century was used to pay for the cost of striking a medal to give to whoever came up with "a new experiment to be propounded and performed every year before the Society," thereby promoting "the advancement of science and useful knowledge."[84] While the accounts of the medallist's virtues can in no way be taken to describe the average Fellow, they are revealing precisely because they delineate the qualities of the ideal Fellow.

In 1748, Folkes singled out Bradley for the two decades of work mentioned above, noting that he had carried on with an:

> ... astonishing consistency. ... [so] scrupulous has he been with regard to the minutest point concerning this

> affair... [that] I dare not determine whether... we should most be affected by the great modesty, the extreme accuracy, the profound judgement, or the surprizing diligence and patience of this great observer.[85]

A year later, Folkes similarly acclaimed John Harrison, the Lincolnshire carpenter-turned-watchmaker who had invented a clock that promised to be so exactly made as to solve the greatest navigational problem of the century, that of accurately establishing longitude. This self-taught man, Folkes rhapsodized, had made a clock that possessed a "degree of exactness that is astonishing and even stupendous." Such a device had been made in consequence of following "an immense number of diligent experiments" performed over nearly two decades.[86] Macclesfield, in 1758, lauded the "long course of experiments" that the optician John Dollond had pursued in his construction of an achromatic lens, which demonstrated his "skill... and exactness." In singling out Dollond—who, like Harrison, was not yet a Fellow at the time of the award of the medal but was an occasional visitor to the Society's meetings—the Royal Society gave "convincing proof of their impartial regard to merit wherever they find it, without confining [their] favours... solely to the Members thereof." In the previous year, Macclesfield had praised Lord Charles Cavendish for the "modesty" with which he made "various useful and instructive experiments" to produce a metallic thermometer; and in the following year he would congratulate John Smeaton for the "remarkable genius" with which he carried out a "considerable number of very accurate experiments" leading to a new understanding of the efficiencies of waterwheels.[87] Each of these men, in other words, was honored not only for his improvement of the stock of knowledge, but also for the qualities of persistence, patience, precision, and politeness that he brought with him to his work.

These qualities were nearly always demonstrated with experiment or observation rather than theory. Even when a theory, however circumscribed, needed to be explained, Fellows rarely reached an understanding of how it worked through an investigation of its pure mathematical underpinnings. The vast majority of Fellows were interested not in the abstractions of mathematics but rather in instrumental, usually mechanical, demonstration.[88] In the 1720s and 1730s J. T. Desaguliers, as Curator, was paid by the Society—in part from the Copley bequest—to produce models that made theories understandable and even entertaining. It was his aim to produce "Machines [that] have been contriv'd to explain and prove experimentally what Sir Isaac Newton has demonstrated mathematically."[89] At the heart of many of these machines was a clockwork mechanism that itself did not need to be explained; rather it bore the explanatory weight.

For example, in 1738 Desaguliers demonstrated a "machine for explaining the tides from the joint operation of the sun and the moon." Five years earlier he had shown "an instrument to represent and imitate the motion of a planet in an elliptic orbit, so as to describe areas about the focus proportional to the times of description"—that is, a machine that made Kepler's law of equal areas in equal times manifest.[90] Likewise, in 1739 the Fellows peered into the innards of a complex watch not to grasp how it worked, but to understand something else; in this instance the variation between mean and true solar time. This remarkable little clockwork universe was opened up by its maker, a Mr. Beridge, to reveal a rotating "wheel, which carried the Index of the true Solar hour [which] was accelerated and retarded at different times of the year by the government of a certain oval Plate which turned round in the space of a year by means of an endless screw."[91] Rather than work through the difficult equation of time, the Fellows could see with their own eyes a curve cut out on a piece of metal and its effect transferred to a dial by a clockwork mechanism. The mechanism itself did not need much explaining; instead it explained other phenomena.

Not all such demonstrations were mechanical. Desaguliers also produced or commissioned instruments that created electrical, magnetic and chemical "performances."[92] Sometimes even magnetic phenomena were given a mechanical explanation. To demonstrate Halley's four-pole theory of magnetic variation, for instance, Captain Hammond "contrived a machine... which [consisted of] a hollow globe having two fixt and two moveable loadstones within it... which answered to the phenomena."[93] Optical phenomena, when appropriate, were also given a mechanical explanation. Thus the newly found aberration of starlight was "represented to the imagination [by a] little instrument ingeniously contrived [to show] how a telescope is to be directed to view an object in the Heavens, as that ray of light may slide down it while the earth is in motion."[94]

Occasionally the desire for mechanical explanation went beyond physical phenomena and reached even into physiology. Dr. Hoadly "shewed the Society a Machine... for explaining the action of several parts concerned in respiration."[95] The Fellows had little patience for models or machines that existed on paper only. John Smeaton, who at the time was apprenticed to a London instrument maker (although he later was to make a career as an engineer), was asked in 1753 to comment on a "machine to measure and determine the way of a ship at sea," which, he acknowledged, was "very ingeniously conceived in the mind of its author," but could have no value until built and extensively tested by a "considerable number of experiments."[96]

While instruments and models explained theories at the Royal Society, their primary role was to produce facts and to uphold the superiority of such facts over hypotheses. The greatest of all authorities—Sir Isaac Newton—had warned against hypotheses in his "Rules of Reasoning in Philosophy," fulminating that "arguements based on induction may not be nullified by hypotheses."[97] This interdiction was taken very seriously and repeated monotonously throughout the eighteenth century at the Society. So well-known was Newton's prejudice that when Monsignor de Molieres applied to become a foreign member in 1729 he assured the Fellows that he was already in his capacity as a member of the Royal Academy "endeavouring in imitation of Sir Isaac Newton to abolish unnecessary hypotheses in physicks" and that he hoped to continue such virtuous behavior as a Fellow of the Royal Society.

When Abraham Trembly sent the Society his remarkable paper on polyps in 1747 he assured them that "we cannot be too circumspect nor too cautious, how we venture to make absolute judgements concerning the nature of things and form to ourselves general rules from those few principles we are at present master of." Similarly, John Dollond, in 1753, dismissed a new optical theory of Euler's, which purported to replace Newton's, because "Mr. Euler has given no reason for his new hypothesis, neither has he produced one experiment in support of it."[98] And John Smeaton, later in the same decade, poured scorn on those who thought they could think their way to a better waterwheel. Anyone who might "imagine... that whenever the same quantity of water descends thro' the same perpendicular space... the natural effective power would be equal" would learn how baseless his assumptions were if only he bothered to check the facts. Indeed, Smeaton was so ingenious as to have conducted his own experiments "without having recourse to theory."[99]

William Mountaine and James Dodson's 1757 collation of 50,000 observations of magnetic variation by Royal Navy and East India Company captains illustrates well the Society's single-minded preference for facts over hypotheses on a global scale. Magnetic variation is the angular difference—measured on the surface of the earth—between true and magnetic north. This quantity itself varies as the observer moves over the globe. In presenting these measurements, which had been made since the beginning of the century, the two men were, they claimed, laying "before the Society nothing but facts without attempting to introduce any hypotheses." So complex were the "magnetic attractions... [of] the great magnet the earth" that the two men judged the phenomena not to be "under the direction of any one general law."[100]

Nor were Mountaine and Dodson the only Fellows to avoid general laws or grand theory in the face of complex phenomena. More typical was a search for relatively modest sets of particular truths: what Desaguliers called "rules that may be reduced to [some] sort of theory"; what the President of the Society, Lord Macclesfield, when discussing Smeaton's experiments on the efficiencies of various waterwheels, labelled "maxims... deductions... observations... corollaries"; and what the London optician John Dollond, in his account of the workings of his new achromatic lens, designated "principles."[101] Books that specialized in grand hypotheses were only occasionally noticed by the Society; and when they were, the reviews were always negative. One work described as "a new system of metaphysics" by Counsellor Cuenz of Neuchâtel was dismissed as "a whole system built upon no firmer a basis than that of arbitrary hypothesis." The reviewer, Dr. Geddes, was thanked by the Society for his "pains" in reading this work.[102] The use of the word "pains" in the Society's Journal Book is rare; more commonly the contributor "was ordered thanks for his account" by the assembled members. Clearly, for most Fellows, the prospect of grappling with metaphysics pained them and the presence of hypotheses, arbitrary or otherwise, horrified them.

Facts, of course, did not merely exist; they had to be established.[103] When some members argued that it might be theoretically possible for a girl to have lost her tongue to a cancerous growth and still be able to talk, other members reminded them sharply that it was desirable

> . . . to have some account of the fact sufficiently well attested, before they endeavor to explain it. For at present they had no account either of the mother's name or the place of her abode, or the names either of the gentlemen or of any other person who had actually seen the girl.[104]

At times the worship of facts could produce uneasy consequences that smacked of saving appearances. On December 17, 1747, a letter from Alexis-Claude Clairaut, one of the few Fellows (nearly all of whom were foreign) who actually practiced mathematical Newtonian philosophy, was read to the Society. In it, Clairaut proposed to modify Newton's inverse square law by adding a new term that accounted for the discrepancy between the unadorned Newtonian theory and the actual "facts" of astronomy. Pushed into a very awkward position, forced to choose between the two idols of English science—namely Sir Isaac Newton or the plain facts—the reviewer of Clairaut's letter lapsed into logical incoherence asserting that:

> . . . facts. . . are all we can build upon in the science of nature. The motions of the heavenly bodies first brought Sir Isaac [to his law]. . . and from the same motions M. Clairaut has learnt that some new modification ought to be superadded to that Law. . . [which will] in no wise interfere with the simplicity of that law or shew it in any other light than as one simple law in fact modifying itself constantly according to the circumstances, and not as a multiplicity of laws imagined occasionally to solve difficulties like Ptolemy's epicycles.[105]

Fortunately for all concerned, it transpired that Clairaut had made an error in his calculations, so that the facts of astronomy no longer implied that Newton was wrong.[106]

Instrument makers were exemplary members and associates of the Royal Society because they could be consistently relied upon to make the instruments that made the facts. Furthermore, they themselves turned to experiment, not hypothesis, to improve their instruments which in turn made more and better facts. Such a virtuous circle greatly pleased the Society and was public proof of its good works. Makers of mathematical and optical instruments such as Graham, Dollond and Ramsden were also respected and honored by the Society for their devotion to another of its prized scientific virtues, accuracy. A passion for exactness—occasionally of theory, but more often of measurement, observation and facts—motivated the activity of many Fellows. Well-made instruments, along with a commitment to disciplined observation, were crucial to the Society's activities because they ensured that the fit between observation and truth was as close as possible. When George Graham compared his observations of the length of a solar eclipse to those of Halley, for instance, he judged his own superior because he knew they could not have erred "from the truth above three seconds."[107]

Foreigners, too, sought to impress the Society by their reverence for accuracy; when the Portuguese ambassador presented a pamphlet on his nation's attempts to fix the longitude of Lisbon by lunar eclipses, he noted that their observations were of "more than ordinary exactness" due to their use of excellent instruments and to the presence of the King himself during observations.[108] If measurements were not sufficiently exact, they were of little use. Desaguliers attacked French measurements that had found the earth to be a prolate sphere in contradiction to Newtonian orthodoxy. The calculations were "just," but the observations themselves had not been "made with that accuracy and exactness which is requisite to determine so small a difference."[109]

While the great majority of Fellows lauded exactitude, a few wondered whether it really could be completely attained. The transits of Venus observations in the 1760s, for example, revealed that even the best prepared observers, using excellent instruments and operating under shared theoretical assumptions, could disagree with each other to a distressing extent. Earlier attempts at a similar enterprise—measuring the shape of the earth—had also given pause to some observers and analysts. The Jesuit natural philosopher Roger Boscovich, for instance, wondered just what theoretical conclusions one could draw from observations that indicated the earth's ellipticity (a measure of how much a spheroid is distorted from the shape of a perfect sphere) ranged from 1 part in 144 to 1 part in 314. James Short, the preeminent maker of reflecting telescopes in London and a constant observer of eclipses, was similarly preoccupied with the question of whether some facts were better than others; he developed a method for averaging observations and discarding those that varied greatly from the mean.[110] Short argued:

> When observations are compared with theories... nothing can be justly inferred against a theory from its disagreement with the observations, unless that disagreement is greater than can be fairly imputed to the imperfections of instruments, and to the unavoidable mistakes of the observers.[111]

When faced with the possibility that facts might not be quite what they seemed—that they might not cast out or confirm hypotheses altogether so simply—instrument makers sought to reduce and even to remove the "imperfections" of the instruments, rather than to develop any new theory of observation or to examine the source and frequency of errors. Similarly, Fellows who used or oversaw the use of instruments did not often consider that observing errors were "natural" and could be effaced by mathematics or by the re-organization of observational techniques. The implication was that those who made errors upon such "perfect" instruments must themselves be morally defective. Thus Nevil Maskelyne, the Astronomer Royal, dismissed his assistant in 1796 for making observations that differed from his own. Such variance—which in the nineteenth century would be normalized out of existence by the personal equation—could, to Maskelyne's mind, only come from a "viciousness" in the assistant. The instruments themselves were above suspicion; they (and their makers) remained "the highest agreed tribunal."[112]

2.2 Cultivating the Tree of Knowledge

Instrument makers and other Fellows tended to live the ideal scientific life at home or in their workshops or in the field rather than at the Royal Society itself. By and large they reported the results of activities undertaken elsewhere to approving Fellows at weekly meetings. What did go on at the Society's premises was the acquisition and judging of books, letters, papers, and specimens of natural or artificial curiosities. It was by submitting these tokens that Fellows and others sought to demonstrate their own devotion to natural knowledge; and it was by keeping and assessing them that active Fellows demonstrated their scientific judgment and discretion. Thus, Simon Schaffer notes, the Society "became a central node of the republic of letters by enticing its 'subjects' to 'submit' their work (as gifts) to its journal. It was this scientific sovereignty the *Transactions* was then able to bestow back as credibility onto its contributors."[113] The actual work of the Society at its premises was to decide what to do with these offerings and what gifts to make to other people and institutions. Books went into its library;

instruments and natural specimens tended to be shown and then taken away by their owners, though the Society did maintain a small collection of mathematical instruments, purchased by Admiralty funds, which it lent out to voyagers such as Captain James Cook.

The Society did, however, keep much paper, taking accounts of its own doings very seriously. In his presidential address to a Copley Medallist with a professional interest in paper (the Copley Medal winner of 1753, and former printer, Benjamin Franklin), Lord Macclesfield spoke of the "illustrious republick" of letters to which the "learned men and philosophers of all nations" belonged and whose members sought to "promote and advance science and usefull knowledge."[114] These literate republicans at the Society spent £500 during the 1720s on making copies of the Council Minute Books and Journal Books from 1660 to the 1720s, and they continued to make copies thereafter.[115] The former volumes minuted the deliberations of the Society's governing body, the Council, while the latter recorded the events of the weekly meetings, including abstracts of letters and papers received. As actual performances of experimental effects became rarer, the office of Curator was abolished in 1744:

> ...the experimental part of Philosophy having begun, even in the early years of the Society, to be cultivated, several gentlemen procured apparatuses to satisfy themselves in private. ... [Soon] a considerable number of Fellows were so well acquainted with the mode of making experiments, that such accomplished Curators have not been found necessary.[116]

Instead, the Society devoted much effort and expense to publishing the work of its Fellows by making sure that the *Philosophical Transactions* appeared continuously and promptly throughout the eighteenth century. Indeed, as Michael Hunter argues, the influence of the *Philosophical Transactions* in some ways "surpassed that of the Society," and the "rise in the bulk of the Society's correspondence and the decline of the corporate experiment were connected, because discussion of the correspondence occupied an increasing amount of the Society's time."[117] The Society also devoted some effort to preserving its own history; a function which remains to this day.[118]

The Society published virtually every paper read before the members at the weekly meetings and by so doing stuck very closely to its founding ideology of registering "nothing but *Histories* and *Relations*" so that others "shall succeed to *change*, to *augment*, to *approve*, to *contradict* them at their discretion."[119] The usefulness of such histories is emphasized by the commercial metaphors that both Thomas Sprat and Francis Bacon used in the seventeenth century to equate the written histories with a capital stock whose earnings will enrich mankind. Bacon speaks favorably of the modern age that is "stored and stocked with infinite experiments and observations," while Sprat asserts that "The *Society* has reduced its principal observations into one *commonstock,* and laid them up in publique *Registers,* to be nakedly transmitted to the next Generation of Men."[120]

The Council of the Society ensured that the *Philosophical Transactions* remained in good health throughout the eighteenth century, first by creating a committee in 1752, to "view the State of the Journal Register and Letter Books, and other Papers of the Society", (taking over the Secretary's task of selecting papers), and then by raising the Society's initial admission fee to five guineas to finance distribution of the journal to all paid-up members in England.[121] Copies were also sent to societies, academies, and universities all over Britain and continental Europe, including those at Oxford, Cambridge, Paris, Berlin, Göttingen, Uppsala, Madrid, Spalding, Peterborough, Bologna, Nuremberg, and Wittenberg.[122] Upon opening the title page, the reader was promised "Some *Account* of the Present Undertakings, Studies, and Labours of the *Ingenious* in Many Considerable Parts of the World."

But just what were the ingenious up to? By turning to a common tree of knowledge—the one that appears in Ephraim Chambers's *Cyclopaedia* of 1728—we can find out which branches flourished at the Royal Society in the eighteenth century and which did not.[123]

Chambers's substantial work, which inspired Denis Diderot's French *Encyclopédie* and earned its author burial in Westminster Abbey, went through several editions in the eighteenth century and was revised towards the end of the century by Abraham Rees. Yet Chambers's 1728 preface was reprinted essentially unchanged by Rees. In particular, Chambers's tree of knowledge (after Bacon), shown in Figure 1, was not touched.

While this tree must have seemed very old-fashioned by the end of the eighteenth century (electricity and magnetism are the most obvious absences), its survival throughout the eighteenth century makes it a useful guide to contemporary divisions of knowledge. That tree had two main trunks: the division between natural and artificial knowledge, or as Chambers also calls it, the "scientifical" and the "technical." The first is what was made of the world through perception or ratiocination using the head; the second is what was made in the world through intervention with technical devices or symbolic manipulation using the hand.

This bifurcation of knowledge corresponds very roughly to eighteenth-century social divisions. The kinds of men who pursued natural knowledge were generally genteel, university trained, and comfortably well-off. They were country gentlemen with landed estates, university dons with endowed chairs, and clerics with ample livings. Even when employing instruments to investigate the natural world, they tended to direct others to make and touch the

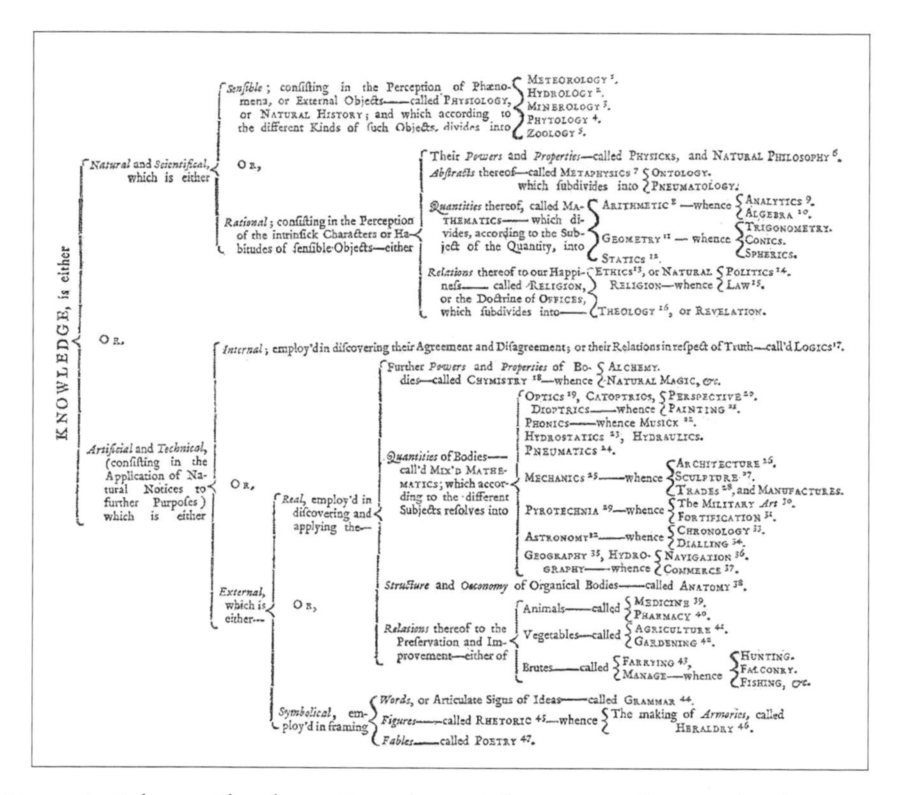

Figure 1: Ephraim Chambers, "Tree of Knowledge," p. ii, Vol. 1, in Chambers 1728; image by permission of the British Library, London.

requisite devices in public, or they amused themselves with instruments in private with no intention of commercial gain.[124]

Those who pursued technical knowledge, in contrast, gained their skills through apprenticeship and by making and directly using instruments themselves. They were chemists, electricians, artists, musicians, opticians, mechanics, tradesmen, military engineers, navigators, apothecaries, surgeons, veterinarians, and gardeners. Such social divisions are not, to be sure, hard and fast. Logicians and other symbolic manipulators (grammarians, rhetoricians, and poets) were often university trained, and the products of their work were symbols, not material objects; while physicians, astronomers, geographers, and alchemists were usually university trained but engaged in artificial pursuits. However, the latter tended either to keep their manual work with instruments private or, if working in public, employed others to manipulate the bodies they were investigating.

The Royal Society, as I shall demonstrate, contained men from both sides of this basic division of knowledge, although not without conflict. But the Society did not claim to take all of knowledge into its purview. Table 4 contains a decade-by-decade content analysis using Chambers's categories of every paper published by the Society for the 1720s to the 1770s; a more detailed year-by-year analysis is given in Appendix A.

Table 4 demonstrates the frequency, if not the quality, of the kinds of knowledge that the Fellows favored. The two great trunks of knowledge—natural and artificial—flourished pretty equally. But only certain branches of those trunks grew vigorously; others were lopped off, or remained rather sickly. While it is not at all surprising to find metaphysics, religion, logic, or poetry absent given the founding rhetoric of the Society, what is very remarkable, in an institution that still basked in the reflected glory of Sir Isaac Newton, is the almost complete absence of pure mathematics (2%) and natural philosophy (3%). Barely one paper a year on either topic graced the Society's journal in the middle decades of the eighteenth century.

As Raymond Cochrane observed many years ago "while the eighteenth-century poets, the world-makers, and the compilers of texts on natural philosophy hailed the god-like Newton as their inspiration, the temper of the age was to be increasingly Baconian."[125]

	NH	MM	MED	ENP	ANA	ATQ	SNP	PM	MSC	Total
1720s	86	80	76	38	40	2	11	12	13	358
1730s	109	94	60	43	29	9	12	6	1	363
1740s	157	84	124	60	58	39	10	8	17	557
1750s	268	82	97	51	24	38	17	11	3	591
1760s	162	172	59	31	17	27	22	9	6	507
1770s	183	94	36	55	28	11	14	19	10	450
Total	965	606	452	278	196	126	86	65	50	2826
	34%	21%	16%	10%	7%	4%	3%	2%	2%	

Table 4: Contents of the *Philosophical Transactions*, 1720 to 1779

ANA anatomy and physiology of animals and plants

ATQ antiquities

ENP experimental natural philosophy: mainly chemical and electrical experiments

MED medicine, surgery, pharmacy, and bills of mortality

MSC miscellaneous

MM mixed mathematics, including: astronomy, geography, mechanics, optics, calculations of annuities, hydrostatics and hydraulics, weights and measures, and pneumatics

NH natural histories of: animals, plants, minerals, waters, the atmosphere, the heavens, manufacturing processes, the earth (including topography), and nations and peoples

SNP speculative or mathematical natural philosophy

PM pure mathematics

Not only is pure mathematics barely visible in the Society's journal, it was virtually never discussed, even in an abstracted form, at meetings; nor was it considered an appropriate topic. For example, in October of 1738 the Society received a mathematical treatise from Clairaut on the figure of the planets. John Machin, Secretary to the Society, was requested to "extract so much as might be proper for reading before the Society." None was.

Less elevated than Clairaut, but no less annoying, were outsiders who sent in unsolicited mathematical puzzles to the corporate body of the Society. When one "proposition in plane geometry, offered by way of a problem," reached the Society, the Fellows declined to deal with such an "improper matter." And even when they could not avoid mathematical work entirely, they refused to be taxed with the nasty details. Thus when an "algebraical tract" by Edmund Waring of Magdalene College, Cambridge, was laid before the Society, only so much of it was read as "did not contain Algebraical calculations."[126] While it is rather hard to imagine what could be left of an algebraical text without the algebra, this is the kind of mathematics that the bulk of the Fellows preferred: that is to say, as little as possible. Pure mathematics—elegant, important and powerful though it was—was considered unfit for polite conversation and remained, silent and occasional, between the covers of the *Philosophical Transactions*.

Natural philosophy—strictly defined by Chambers as "the perception of the intrinsic characters or habitudes of sensible objects" or by another contemporary encyclopedist, John Harris, as an investigation into "the Natures and Causes of Things"—did not fare any better than pure mathematics.[127] Due to the Society's intense dislike of any form of speculation, there were exceptionally few discussions about the natures and causes of such phenomena as gravity, mechanical force, electricity and magnetism. Thus the kind of speculation into the properties of bodies that one finds in Newton's "Queries" to his *Opticks*, or the kind of organized account of the system of the world that he gave in his *Principia*, were more honored in the breach than the observance at the Society he had presided over.

Scarcely less surprising is the rather modest amount of experimentation reported in the Society's journal. The science that exemplified experiment in Chambers's day was chemistry, which he defines in the main body of his *Cyclopaedia* as having as its object "all sensible bodies, capable of being contained in vessels." Chambers gives this central science a category of its own, but, like d'Alembert in his encyclopedic tree of knowledge a generation later, makes no mention at all of electricity which, along with the less common study of magnetism, would become a more popular experimental subject later in the century. Despite the emphasis on experimental philosophy voiced by the early promoters of the Royal Society, and even despite the explicit encouragement of experimentation by the annual awarding of the Copley Medal, experiments were not particularly common, accounting for only 8% to 13% of papers during each decade from the 1720s to the 1770s and averaging 10% for the entire period.[128]

My finding is in good agreement with other investigations. Charles Bazerman estimates that 10% of the papers between 1665 and 1800 actually describe experiments,[129]. Marie Boas Hall has similarly demonstrated that experimentation, initially a major concern of the Society, was practiced with ever-decreasing frequency in the decades after the Society's founding and on into the eighteenth century.[130]

If the Fellows only very rarely practiced pure mathematics or speculated upon the underlying causes of things and not much more frequently investigated the world by experimentation in laboratories, what then did they do? One contributor to the Society's journal gave an answer in 1759: "the English way of philosophising is not to sit down in one's study, and form an hypothesis, and then strive to wrest all nature to it; but to look abroad into the world, and see how nature works; and then to build upon certain matters of fact."[131] Natural history and antiquities (38%); medicine, surgery and anatomy (23%); and mixed mathematics (21%); got the Fellows off their chairs and out of their studies: walking through fields and over mountains, riding to visit patients, and sailing the world's oceans. Only by living an active life could the natural historian or the physician or the surgeon gather histories of animals, vegetables, minerals, waters, airs, and peoples in their regular and irregular appearances.[132]

Mixed mathematicians, too, had to leave their books behind in order to practice the important components of their discipline, which included astronomy, geography, mechanics, optics, geodesy, metrology, navigation, surveying, fortification, and architecture. Even the most sedentary mixed mathematicians—astronomers—had to travel at times to other parts of the globe to make observations. Geography, navigation, surveying, and geodesy could be nothing more than idle speculation if its practitioners did not find and map new lands, sail across oceans, mark out boundary lines, and measure the curvature of the earth. Likewise, an architect or engineer could scarcely expect much business unless he travelled to study existing buildings and machines and to oversee the construction of new ones. Only instrument makers, the men whose instruments underlay all this worldly measurement, did not need to travel outside of London. They made and improved their mathematical and optical instruments, as I shall show in the next chapter, by private experiments in their workshops, discussions with other experts, and lessons from books and lectures.

Mixed mathematics no longer exists today, but it was a vast and important discipline in the eighteenth century and the utility and quantifiability of its subjects were very much to the Royal Society's empirical taste. Unlike pure mathematics, which dealt with quantity itself detached from the properties of any particular body, mixed mathematics measured the quantities of particular, real, bodies—whether they were moons, mirrors, or machines—and put them to a practical purpose.[133] Since this was the discipline in which mathematical and optical instrument makers most frequently displayed their talents, I have in Table 5 considered in more careful detail the distribution of mixed mathematics papers in the *Philosophical Transactions*.

	Astronomy	*Geography*	*Mechanics*	*Other*	*Total*
1720s	32	38	6	4	80
1730s	23	45	19	7	94
1740s	40	19	14	12	85
1750s	30	18	15	19	82
1760s	106	36	11	19	172
1770s	33	35	8	18	94
Total	264	191	73	79	607
	43%	31%	12%	13%	

Table 5: Mixed Mathematics in the *Philosophical Transactions,* 1720 to 1779

Astronomy natural histories of the heavens, mapping of stars, other positional observations not for geographic purposes

Geography cartography, navigation, surveying, shape of the Earth, and observations to fix latitude and longitude

Mechanics ballistics and accounts of machines

The great majority of these papers give accounts of astronomical observations, yet they cannot always be placed into the subcategory of astronomy, since most of their observations served, either directly or indirectly, to fix latitude and longitude and thus more properly belong to another part of mixed mathematics, geography. In fact, in this very practical sense, astronomy was almost entirely subordinate to geography in the eighteenth century; observatories were explicitly founded and funded to advance navigation and surveying, not to test Cartesian or Newtonian theories of the cosmos. Even when observers seemed to be sharpening their astronomical skills and enjoying the pleasure of accurately observing relatively rare celestial events, the practical aims of geography were never far behind. An observation of a solar eclipse or transit at a single given location, for example, could do little more than provide an astronomical spectacle for its observers and provide substantiation for theories predicting its occurrence. When made in concert with other observations around the globe, however, the event served to fix the relative distance between the observers (longitude), a prime task of geography.

Thus, for example, before M. De L'Isle set out for Russia to observe the 1739 transit of Mercury from the banks of the Ob River near 67 degrees north, in order to give his isolated observation a global meaning, he first wrote to the Royal Society requesting that an English astronomer might observe the same event so that he could then fix "the true longitude of the place... in order to render the observation of general service."[134]

Where astronomical observations were gathered explicitly for fixing latitude and longitude, I have placed them in the sub-category of geography along with papers that discuss cartography, the shape of the earth, and navigation. Remaining in the sub-category of astronomy are only those papers where geographical concerns are not at all explicit or are completely absent. Examples of the former include repeated observations at locations whose longitude is well known; the latter include the great number of papers in the 1760s dealing with the two transits of Venus, which served to answer the purely cosmological question of the absolute size of the solar system. The final two mixed-mathematics sub-categories are less problematic, if much more scattered. The first—mechanics—includes papers that consider the powers of machines or the mechanics of projectiles. The second—miscellaneous—deals with optics, hydrostatics, hydraulics, pneumatics, and calculations of various mathematical tables, including logarithms and annuities. Papers hypothesizing on causes of effects are almost completely absent.

Mixed mathematics remained a remarkably constant concern of the Fellows throughout the eighteenth century. If we put to one side the two transits of Venus in 1761 and 1769, which generated a huge number of astronomical papers in the 1760s, the Society published about ninety papers per decade on the subject with little variation. More dramatic changes occur, however, in the other two main branches of activity at the Society. Natural history became increasingly dominant throughout the century, almost doubling in quantity from the 1730s to the 1740s and maintaining numbers thereafter of close to two hundred papers a decade. The anomalous number of papers in the 1750s is due to a series of extraordinary natural events, in this case earthquakes that shook much of Europe, including one that destroyed Lisbon in 1756. In stark contrast to the rude health of natural history, papers on medicine, surgery, and anatomy withered away from nearly two hundred a decade in the 1740s to barely seventy in the 1770s.

No single hypothesis explains why these four main branches of the tree of knowledge were cultivated at the Royal Society in the way that they were. Take utility. In a commercial, maritime, imperial nation such as eighteenth-century Britain, a paper that helps fix the longitude of an important foreign port or describes a new shipboard ventilating machine or telescope lens is obviously useful to the nation. In the journal of a Society whose founding rhetoric exalts the virtue of utility, the appearance of such publications is not surprising. But many other papers—including the very great majority of those found on the massive trunk of natural knowledge—are almost deliberately useless. Papers that describe two-headed calves, demonstrate entertaining electrical effects, or speculate upon the vacuum in an air pump cannot readily be given a utilitarian explanation.[135] Try then "gentility." One could argue that such activities and their descriptions were socially acceptable for gentlemen to engage in, precisely because they were not tainted by the pragmatic and prosaic interests of trade and commerce. But this explanation cannot then work for the useful papers. Thus "utility" and "gentility" are in very many instances at cross purposes with each other.

In fact, this conflict between the production of utilitarian and genteel knowledge is roughly inscribed into Chambers's tree itself. The men who produced natural knowledge did not—with a few important exceptions—depend upon the production of that knowledge for their primary living.[136] To put it bluntly, they did not make money out of science. Nor did they identify themselves as naturalists. They were, first and foremost, gentlemen. Contrariwise, artificial knowledge was, by and large, made by men—physicians, surgeons, apothecaries, navigators, engineers, mechanics, opticians, painters, and musicians—who identified with the branches of knowledge they made a living from.

These utilitarian, technically facile men may have had a lower status at the Society, and at times of tension they were put in their place, but they were not excluded.[137] Quite to the contrary, they were among the most eminent, energetic, and internationally famous of its Fellows. The skills that brought them success in their working lives—a delicacy of touch, mechanical brilliance, and a taste for geometry—were displayed to an approving Royal Society. However, these appropriate behaviors were not alone sufficient to make such men Fellows. They also needed certain social characteristics that fitted them to be members of the particular London club that met in Crane Court.

2.3 Clubbability

Like other London clubs, the Royal Society conformed to contemporary notions of sociability among men of many ranks and intellects and "provided a setting in which private friendship and formal organization were judiciously balanced. The inhibitions imposed by the hierarchical bonds of traditional society [were] diminished."[138] This was not to say, however, that anybody could belong to the Society. Men of high social status joined easily, those of middling status less so, and those of low social class not at all.

In the previous century, as Michael Hunter has convincingly shown, the culture of the virtuosi completely dominated the Royal Society to the virtual exclusion of those with lesser social and intellectual standing. For example, almost no surgeons, apothecaries, schoolteachers, or instrument makers belonged in the seventeenth century.[139] But by the early eighteenth century men like this did start to appear in the Society's ranks. The success of the Newtonian philosophy, as explicated in texts, lectures, and sermons among various sorts of men, helped turn even those who had not had a gentleman's university education into plausible candidates for election to the Society.

All the same, instrument makers had to walk a very fine line between commercial success and social approbation. None of those elected was single-minded in his pursuit of wealth nor ever explicitly stated he sought membership in the Society to improve his business opportunities. Edmond Halley recommended George Graham to John Harrison telling him "Mr. Graham was a very honest Man." Graham went on to astonish both contemporaries and subsequent historians by freely assisting the notoriously paranoid Harrison with his longitude-finding chronometer but making no claims for the Longitude Prize. John Dollond was praised for being "studious and philosophic." And Jesse Ramsden was said to have lived a life "of extreme frugality," leaving what little money he had to his workmen after his death.[140]

Those who behaved differently were excluded. Benjamin Martin, a maker of excellent reflecting telescopes and a relentless self-promoter, wrote in 1741 to the President of the Society, Sir Hans Sloane:

> I am constantly asked, *If I am a Fellow of the R. Society*? And as I constantly find it no small Disadvantage to say,

> No. I have therefore been advised and prevailed upon to apply for that Honor, which would give so great a Sanction, and be of such Advantage to my Profession.

This blunt approach was completely disastrous and Martin never became F.R.S. nor had much further to do with the Society.[141]

Another maker of excellent telescopes, even more successful in his business dealings than Martin, was the optical instrument-maker Peter Dollond. His father (John Dollond), his brother-in-law (Jesse Ramsden), and his nephew (George Dollond), were all Fellows, but he never became one. While there is no evidence that he ever applied, his membership in the American Philosophical Society, family links to the Royal Society, attendance at meetings of the Royal Society under the patronage of Nevil Maskelyne or John Smeaton, publication in the*Philosophical Transactions*, and construction of important and innovative telescopes for major observatories in Britain and the rest of Europe would have made him a likely candidate. However, he was not modest; rather he was so aggressive as to take on the entire London optical community in a series of successful lawsuits in the 1760s, clearing the way to significant commercial gain. This grossly self-interested action—which I shall investigate more closely in Chapter Five—disqualified him for membership at the Royal Society in the eyes of many Fellows.

Another disqualification was insufficient reputation as a maker of innovative instruments or as a practitioner of the Society's methods. Thus the prominent London optician John Cuff, who had the support of several eminent Fellows (among them Lord Egmont and two future Presidents of the Society, Folkes and West) was rejected on April 19, 1744, because, in the standard and uninformative language used in these cases, "he was not sufficiently well known to a proper number of members who happened to be present."[142] He continued, however, to attend meetings from time to time during the next two decades. Cuff was rejected not because of his social class alone. He was sufficiently wealthy and respected to be elected Master of the Spectaclemakers' Company in 1748 and to have his portrait painted by Zoffany.[143] But he did not make the kinds of contributions to mixed mathematics or natural philosophy that merited election, and he contributed no papers to the *Philosophical Transactions*.

While election was never certain for instrument makers or indeed for most other classes of men, it was automatic for royalty, aristocracy, Privy Councillors, and foreign diplomats. Unlike other prospective members, who had to wait 10 meetings between proposal and possible election, such men could be proposed and automatically elected at the same meeting.[144] Yet despite such marks of social differentiation, other practices indicate that a kind of equality generally existed within, if not without, the walls of the Society. Every member, in imitation of the House of Commons, had a vote of equal value. The most important elections were the occasional ones for the President of the Society (who governed until he resigned or died in office) and the annual ones for the governing body, the Council. The Council and the President ruled completely without royal interference despite operating under a royal charter.[145]

Funding—or even indications of interest from the sovereign, following the precedent set by the Society's first patron, Charles II—was extremely rare. George I was still not patron of the Society when Hans Sloane, the new President, came to power in 1727. Newton so hated Leibniz (who had the support of the Hanoverian George I) that he did not pursue the formal patronage of the King upon his succession to the throne in 1714 lest the new King "endeavor to moderate" the Royal Society's and Newton's handling of Leibniz and "perhaps even to find a place for his philosopher in the direction of the Society's affairs."[146] Resolving this embarrassing, and potentially dangerous, state of affairs turned out to be not entirely straightforward.

In a most unusual meeting on May 11 of that year, Sloane announced that the King had agreed to become patron and that the Society should consider in what manner to ask the King's son, the Prince of Wales, to do so as well. Some members proposed voting on the wording of the request, while others, who were in the majority, strongly objected to any hint of democratic procedures in "putting questions of this nature to a ballot."[147] The latter party was successful and the Society unanimously agreed, after much debate, not to put the question.

A related discussion, also involving issues of loyalty and the right of a freeborn Englishman to express his opinion among his fellows, took place during a meeting two weeks later. John Byrom, a Tory and the leader of the attack on Sloane's new presidency later in the year, had spoken vigorously at the meeting a fortnight previously "upon a question of free debate, whereof he was heard by the Society without censure, not having transgressed the bounds of that liberty of speech which every member has an undoubted right to use." At that time he had noticed a visitor who took notes "in an unusual manner... [and] wrote down his name and was very solicitous to be informed about the right spelling of it" during his speech. The visitor was Mr. Ahlers, a surgeon at the Court and, as Byrom went on to note darkly, a "foreigner" who might not understand, or, even worse, might misrepresent to the Court, the "privilege and liberty of speech" enjoyed by all Fellows. Several other members supported Byrom and demanded an explanation from Ahlers.

At this stage the affair dissolved into bathos when Ahlers explained, to the apparent satisfaction of the members, that he had trouble remembering or spelling people's names and wrote them down to assist his memory.[148] The dispute spilled over into a challenge to Sloane's presidency, initially from the Tory faction within the Society and then from others who objected to Sloane on personal and intellectual grounds. The challengers were easily defeated in November of 1727.[149]

Despite the resolution of the May and November disputes, the episodes of 1727 demonstrate that many Fellows had strong feelings about the privileges of free speech attendant upon their own social standing, the protection of the law, and the proper exercise of liberty within a Society of putative equals. Of course, as in the House of Commons, some members were "more equal than others." Nonetheless, traces of a presumed parity among members persisted. When not given specific titles, members were usually referred to as "gentlemen": for example, most election certificates opened with the phrase, "a gentleman well versed in... most branches of curious and natural learning," and most meetings referred even to visitors as "gentlemen [who] had leave to be present." Occasionally members—possibly echoing the usages of Quakers or Freemasons—referred to each other as "brothers."[150]

High social rank (with the exceptions already noted above) could not guarantee admission to the Royal Society; nor, however, was modest social rank an impassable barrier. For example, John Wreden Esq. (surgeon to the Prince of Wales) and James Gambier Esq. (a man who, according to his recommenders, was "well versed in natural knowledge") were both rejected on February 13, 1734/35, as was John Locke Esq., a lawyer from the Inner Temple also rather vaguely said to be "well versed in natural knowledge," some two months later on March 27, 1735. At the latter meeting, however, the rather more plainly styled and professioned Mr. Martin Clare, a schoolmaster from Soho Square, who came with the specific recommendation of being "a good mathematician, well skilled both in natural and experimental philosophy, and a great promoter of the same," was elected.[151]

Any sample of the rather few actual rejections is, to be sure, inevitably skewed, since only those men who thought they were the sorts of people who should be Fellows were likely to apply. However, the sample does demonstrate that being a gentleman vaguely interested in natural philosophy and possessing certain social and intellectual qualities was not always sufficient for election, while being a schoolteacher who earned a living from mathematics or an instrument maker who worked with his hands was not necessarily an obstacle.

A better indication of the reasonably varied social composition of the Royal Society in the middle years of the eighteenth century is given in Table 6, which I have compiled by examining election certificates for every fifth year, beginning in 1735 and ending in 1780.[152] To get a sense of changes in social composition over time, I have, in Table 7, compared Table 6 with Michael Hunter's similar compilation for 1660-1685, the first twenty-five years of the Society.[153] I have not considered foreigners in my sample and have removed them from Hunter's table and scaled his numbers appropriately. Where Hunter's and my categories have not coincided I have added two or more together. For example, the category for professional scholar works less well in the eighteenth century and I have put such people in the appropriate university degree category, which is nearly always clerical. Thus Hunter's categories of divine and professional scholar are combined to compare with my combined categories of cleric, bishop, and schoolteacher.

At first glance, the most striking aspect of Table 7 is the similarity it shows in the membership between the two periods; for instance, the respective percentages of important personages (peers and high government officials), gentlemen, lawyers, physicians, and clerics all remained fairly stable, together comprising about 91% of the membership between 1660-1685 and about 87% between 1735-1780. Yet Table 7 also provides evidence of a certain, if by no

	'35	'40	'45	'50	'55	'60	'65	'70	'75	'80	Total	
Peer	5	5	5	3	2	4	2	1		6	33	20%
Bishop				5			1				2	1%
Baronet or Knight	1		2	5			3		2	1	9	5%
Esquire or Gentleman	2	3	2	5	1	4	4	3	3	2	25	15%
Military Official				5				1	3	2	7	4%
Government Official	1	2	1	5				2	2	1	10	6%
Lawyer Judge	1		3	5	1		1			1	8	5%
Physician		2	2	5	1	1	5	6	3	1	27	16%
Cleric	1	2		5	5	2	2	4	2	1	21	12%
Surgeon or Apothecary	3		1	5	1		1	3	4	1	16	9%
Schoolteacher	1	1	1	5	1						5	3%
Other			1	5			1		3		6	5%
Total	15	15	20	18	12	11	20	20	22	16	169	
Foreigner	7	13	5	12	11	13	9	3	2	5	80	

Table 6: Social background of Fellows of the Royal Society from an every-fifth-year sample of the Royal Society's election certificates

means radical, degree of expansion in the social class of the membership, demonstrated in particular by the appearance of surgeons, apothecaries, and schoolteachers, who did not belong to the Society in the seventeenth century but who by the mid-eighteenth century comprised about 9% of the membership.

Margaret d'Espinasse was the first to note the disappearance of wealthy merchants from the Society's membership but her inference that the Society had, as a consequence, narrowed its intellectual interests and social membership by the eighteenth century is not upheld by the data in this chapter.[154] Although instrument makers were too few in number to warrant a separate category of their own (most would be placed under "Other"), their admission from the 1720s onwards provides further evidence of the Society's social expansion.

Alongside the Fellows in Table 6 who belonged to three of the oldest professions (law, theology, and medicine) or to the aristocracy and landed gentry, were members who are not so easily classified. As the table indicates, two-fifths of the members were gentlemen by virtue of their social position alone (from peers down to gentlemen); another two-fifths belonged to established, genteel professions (military officers, lawyers, physicians, and clerics); and the remaining members undertook a variety of occupations. There was much variation of status within each grouping. The grandest of military officers and government officials purchased their office and were often the sons of the aristocracy and landed gentry, while some attorneys and clerics were not much better off than a prosperous master carpenter or grocer.

1660 to 1699			*1735 to 1780*
Aristocrat	14%	20%	Peer
Courtier or Civil Servant	20%	15%	Baronet, Knight, Military or Government Official
Gentleman	15%	15%	Gentleman
Lawyer	4%	5%	Lawyer
Physician	16%	16%	Physician
Surgeon or Apothecary		9%	Surgeon or Apothecary
Divine, Scholar or Writer	20%	16%	Cleric, Bishop, or Schoolteacher
Merchant	7%		
Unknown	3%	3%	Other

Table 7: Comparison between the Membership of the Royal Society in 1660–1699 (Hunter, 1976) and 1735–1780 (Table 6)

Whether or not members of this last class of men, which included among others surgeons, apothecaries, and schoolteachers, would have called themselves gentlemen, they certainly had to be men of polish and politeness to succeed in their businesses or newly forming professions.[155] They were not landed gentry, but they aspired to a genteel lifestyle. They cultivated and enjoyed a national and international fame that both sold their wares and increased their stature, since society in mid-eighteenth-century Britain was "so flexible, the economy was growing at [so] many different points... that all sectors of society supplied the men of enterprise."[156] A burgeoning consumption of luxury goods blurred the lines among the landed gentry, the ancient professions, and others with a claim to gentility. Commercial activity was not looked down upon as being necessarily un-gentlemanly in the eighteenth century; indeed, both Adam Smith and David Hume argued that it was the basis of the highest possible stage of civilization.[157] Commerce brought wealth and an ever-increasing exchange of objects and ideas. Wealth and the division of labor in turn brought leisure and the opportunity to cultivate politeness, learning and religion. Thus in eighteenth-century England, the gap between polite society and commerce was not an unbridgeable one.

However, that gap did still exist. As I have already noted, instrument makers had to be very careful as they travelled between the worlds of commerce and polite learning. As the eighteenth century drew to a close that journey became ever more precarious. In 1784, a dispute broke out when the President of the Royal Society, Sir Joseph Banks, demanded that the assistant secretary for foreign correspondence—Joseph Hutton, professor of mathematics at the Woolwich military academy—must live in central London, near to the Society and more especially to Banks himself. Hutton resigned rather than move away from his primary job. He and his friends interpreted Banks's move as an attack on mixed mathematicians—most of whom, as did instrument makers, had to work for a living—and the controversy broadened into a furious debate about the proper nature of the Society in general and the role of mixed mathematics in particular.[158] Banks himself thought that mathematics was "little more than a tool with which other sciences are hewd into form." His supporters agreed that "it was not sufficient to be a Mathematician, to be a Fellow of [the] Society; there were. . . requisite social qualities, the lack of which might make a man unfit, however competent he might be in learning."[159]

While the mathematicians were initially victorious when votes were held among the regularly attending active members, they were crushed when Banks marshalled the full membership of the Society against them. The mathematicians lost because, as John Heilbron notes in his perceptive analysis of the episode, mathematics "had about it an aroma of trade and tedium that did not recommend it to gentlemen."[160] They also lost for the elementary reason that they were very much a minority at the Society. They do not even warrant a category of their own in Table 6 and are to be found scattered among the small categories of military officers, government officials, and schoolteachers (13%). If we put to one side physicians, surgeons, and apothecaries (25%) who, while relying upon their expertise for a living, were a complicated mixture of genteel professionals and tradesmen, the Society's remaining domestic members (68%) were not dependent upon science for a living and were mostly gentlemen. Indeed, this figure is understated since many of the military officers and government officials were also gentlemen who held office by virtue of their social position and connections.

But the social fissure at the Society between those who were primarily dependent for their living by working at science and those who were independent gentlemen—or, for that matter, between those whose tastes ran more to natural knowledge and those who pursued artificial or technical knowledge—was not the only chasm that could open up. The Society was also split between active members who regularly attended meetings or published in the *Philosophical Transactions* and those inactive members who did neither. The largest block of members, the silent majority of the Society that Banks mobilized in 1784, consisted of the inactive independent naturalists who, in concert with other naturalists and independent gentlemen, roundly defeated the active but dependent producers of technical knowledge.

Men whose interests straddled natural and artificial knowledge were caught in an uncomfortable position. For example, Henry Cavendish was a brilliant exponent of natural philosophy and pure mathematics, as well as chemistry, electricity, and mixed mathematics, and he admired and socialized with the instrument makers upon whom his experimental and mixed mathematical work depended. He was also so great a gentleman—the grandson of two dukes and vastly Banks's superior in both social position and wealth—that he could have done whatever he pleased. He wished initially not to be involved in the controversy; but when he could no longer avoid it, he cleaved to his class interest and supported the naturalists and Banks over the mixed mathematicians.[161]

If they were in such a minority and so easily defeated, why were the mixed mathematicians not cast out? As Table 4 demonstrates, a Royal Society shorn of mixed mathematics would lose one-fifth of its knowledge production and some of its most internationally prestigious members. If we look at Chambers's tree of knowledge, we can see that the Society could not lop off so massive a branch and remain credible abroad. Once bereft also of astronomers, surveyors, navigators, cartographers, military engineers, actuaries, opticians, and mathematical instrument makers, a Society that already had practically no mathematicians or natural philosophers would be the laughingstock of scientific Europe. And if the mixed mathematicians left or were pushed out, would the chemists and electricians be far behind? Faced with this alarming possibility, the silent majority retreated to their country estates and were never again deployed by Banks.

Indeed, a decade later, Banks was forced to acknowledge in his Copley Medal address one of the most eminent of the mixed mathematicians, Jesse Ramsden. However, as we shall see in the Epilogue, the gentlemen of the Royal Society resented this dependence upon technicians and Ramsden's triumph was to be one of the last for the mathematical instrument makers. The fissures at the Society persisted well into the nineteenth century and some groups were eventually pushed out. Thus inactive Fellows alongside medical Fellows were decreasingly elected in the 1830s and 1840s and had pretty much disappeared by the 1860s. One basic split—in faint echo of 1784—remains to this day: the Society's presidency alternates between scientists from the physical and mathematical sciences on the one hand and the biological and earth sciences on other and the Society publishes the *Philosophical Transactions* in a Series A and a Series B corresponding to the same division.[162]

Chapter 3

Measurement and Experiment

Instrument makers were valued members of the Royal Society because their overall approach to science suited them to become Fellows. The most important components of that approach included careful observation, mechanical ingenuity, factuality, accuracy, and the pursuit of perfection by means of the experimental philosophy, all of which characterized both the instrument-makers' behavior and the scientific ideals of the Society itself. Pervading these behaviors was an instrumentalism which instrument makers, and many other Fellows of the Society, believed led to advances in natural philosophy through the improvement of instruments. Historians have also made this argument. Allan Chapman points out that theoretical advances in astronomy were impossible without prior instrumental advances, while John Heilbron notes that it was "the study and perfection of instruments... not the requirements of high theory" which "provoked the quantification of the physical sciences during the two decades on either side of 1800."[163]

Instrument makers used or provided instruments for two of the Society's main activities: mixed mathematical measurement and natural philosophical experiment. They enriched the former by providing, and sometimes themselves using, instruments of unprecedented accuracy to make astronomical, geodetical, and geographical measurements. These not only benefited the imperial interests of Great Britain but also led to unexpected discoveries and the settling of vexing controversies in natural philosophy, helping to reveal the immense scale of the cosmos in the 1720s, the shape of the earth in the 1730s, and the geographic outlines of continents and islands in the 1760s and 1770s. In addition, as part of their unceasing attempt to improve the accuracy of their instruments, they conducted careful and lengthy experiments in magnetic and optical phenomena in the 1720s and 1750s respectively.

3.1 The Aberration of Starlight

Before setting out on such ambitious investigations, instrument makers had first to establish their credentials as trustworthy observers. They did so by comparing themselves to, or associating themselves with, men of high reputation. In astronomy, the first exemplar in the nation was the Astronomer Royal who, in 1722, was Edmond Halley. His observations of an eclipse of the sun on November 27 of that year, made in Greenwich, were listed first in the *Philosophical Transactions*; George Graham's, in central London, followed. Ambitious and skilled clock and watchmakers such as Graham were expert at observing transits of stars to check the going of their timepieces against that most perfect clock, the spinning earth. Graham thus assures the reader that he had done just that, having made "very correct Observations both of the Sun and Stars, the 26, 27, and 28th, for determining the exact Time by my Clock." His observations of the beginning and the end of the eclipse, which lasted more than two hours, differed by less than 30 seconds from those of Halley.[164] This was the last time (with one exception) that George Graham's observations followed anyone's. In seven of his eight subsequent astronomical papers that are grouped with those of others who observed the same event, Graham's observations come first. By 1736, having become famous for equipping the Royal Greenwich Observatory and making a range of important global measurements, he had even displaced Halley from first place. Perhaps one reason why Halley allowed Graham's observations eventually to supersede his own was Graham's part in the discovery of the aberration of starlight. This remarkable phenomenon, although in retrospect a necessary consequence of the Copernican system was, as Halley noted, "never suspected from the beginning of Astronomy."[165]

By the 1720s, nearly two centuries after the death of Copernicus, another phenomenon that was a necessary and very well-accepted consequence of the Copernican system—stellar parallax—had not yet been proved by observation. Stellar parallax is a very tiny effect that makes a given star seem to move slightly against the general stellar background throughout the year as the earth orbits the sun. To explain away the lack of an observed parallax, astronomers argued in the early eighteenth century that the parallax was too tiny an effect to be seen with current instruments and hence the closest star was very distant indeed from the sun. This explanation turned out to be correct, but the stellar parallax was not directly observed until Friedrich Bessel's work more than a hundred years later.

The solution to the problem could only lie in the construction of yet more accurate instruments; and George Graham was the man to whom James Bradley turned to construct such a device that both men hoped would finally demonstrate the parallax. Graham made a zenith sector that was, as Bradley put it, "exact and well-contrived."[166] The zenith is the point directly over the observer's head, and the meridian is the plane that passes through the zenith and the earth's axis. The sun passes through the meridian at midday, while all other stars pass through the meridian at a slightly different time each day, returning to the same time after a year. One December night in 1725, while looking for the parallax, Bradley noticed that a star crossing the meridian near the zenith was 3" more southerly than he had expected. Although 3" was a very small amount—Brahe's best observations, at the beginning of the previous century, were some sixty times less accurate, being good only to a few minutes—Bradley believed that Graham's instrument was at least this accurate so he continued for an entire year to measure the angle at which the star passed through the meridian. Since the star was near the zenith, errors due to the refractive effect of the atmosphere were not significant. Bradley also took into account the precession of the equinoxes when calculating the change in the star's position. During that year the star first continued in a southerly direction, then moved northerly, and finally returned to its original position in December of 1726, which Bradley took as "sufficient Proof that the Instrument had not been the Cause of this apparent Motion of the Star."

Bradley had two reasons to assume that the completely unexpected phenomenon was not an instrumental artifact. Firstly, the returning of the star to its original position after exactly one year implied that the phenomenon was most likely caused by the annual motion of the earth. Secondly, and perhaps more compellingly to Bradley, Graham's sector produced errors much smaller than the phenomenon itself; hence "we can be well assur'd from the perfections of our instruments, of the limits wherein the error of our observations are contained, [and] we need not hesitate to ascribe such apparent changes as manifestly exceed those limits to some other causes."[167] Confident in the perfection of Graham's instrument and, in good Royal Society style, putting "aside all thoughts...about the Cause," Bradley set out to acquire the "proper Means to determine more precisely" what exactly the phenomenon was. The "proper Means" was a new and even more precise zenith sector made by George Graham: a "more accurate instrument" than any that had preceded it.[168] It was twelve-and-a-half-feet long and, while Graham did not build it all himself, he certainly carried out the most delicate work, dividing the arc and setting the micrometer screw and the crosshairs of the telescope. Bradley believed that this instrument was accurate down to 1/2"; he selected 12 stars near the zenith and settled down to another year's observation before endeavoring to "find out the Cause."[169]

Bradley hypothesized that the phenomenon he had observed could only be caused by the relative motion of the earth with respect to the stars. If a telescope on the earth were pointing directly overhead at a star and the earth were moving past the star, then the light entering the top of the telescope would not reach the bottom, since the telescope would have moved past the star. On the other hand, if the telescope were inclined slightly from the vertical by an angle equal to the ratio of the earth's speed to the speed of light, the starlight entering the top of the telescope would get to the eyepiece in the bottom as the earth moved past the star. Bradley supposedly solved the puzzle, after he had his observations in hand, when on a boating trip he noticed that the flag on the boat's mast moved as the boat went about, even though the wind direction had not changed. From this he concluded that the relative motion of the boat brought about an apparent change of the wind direction as measured by the flag. Therefore, by analogy, the relative motion of the earth brought about an apparent change in the star's position as measured by the telescope. Unfortunately, this charming story is found nowhere told by Bradley himself.[170]

Popularizers of astronomy in the eighteenth century used another analogy, equally suited to the British, namely, likening the phenomenon to walking in the rain with an umbrella: the faster you walk, the more you must incline your umbrella to avoid getting wet.

Likewise, the faster the earth moved, the more inclined the telescope would have to be to collect the light in its eyepiece. Bradley observed that, over a year, the star traced out a small ellipse around its expected position. From the size of this ellipse and the star's latitude, he concluded that the sun's light took 8 minutes 13 seconds to reach the earth.[171] Halley regarded the "curious observations and reflections" that lay behind the establishment of the aberration of starlight to be among the most remarkable events in the history of astronomy.[172] Not one, he stated, but "three Grand Doctrines in the Modern Astronomy do receive a Great Light and Confirmation from this one Single Motion of the Stars Light Viz: the motion of the earth, the motion of light, [and] the immense distance of the Stars [from the sun]."[173] Furthermore, this was the very best kind of discovery: it had been made "accidentally," which meant that the observer was not trying to prove any pre-existing hypothesis. Nature had been caught unawares by a newly precise instrument.[174]

Bradley clearly believed that natural philosophy could be improved by increases in observational accuracy. Twenty years later, in his paper of 1748 demonstrating the subtle phenomenon of the nutation of the moon's axis, he would more explicitly espouse this conviction: Kepler had relied on Brahe, Newton could not have philosophized without the invention of clocks and telescopes, therefore theory is indebted to practice for many of its inductions. All these instances, Bradley concludes, "point out to us the great Advantage of cultivating in *this* as well as every other Branch of Natural Knowledge... a regular Series of Observations and Experiments."[175] Bradley gave a glowing testimonial to his friend, the man "who above all others, has most contributed to the improvement" of astronomical instruments "our worthy Member *George Graham*." Bradley went on to note his dependence upon Graham:

> ... if my own Endeavours have, in any respect, been effectual to the Advancement of Astronomy; it has principally been owing to the Advice and Assistance given me by [Graham]... whose great Skill and Judgement in Mechanicks, join'd with a complete and practical Knowledge of the Uses of Astronomical Instruments, enable him to contrive and execute them in the most perfect manner.[176]

This is an extraordinary encomium from a gentleman and an Oxford professor to an artisan and a London shopkeeper. Not only was Bradley reliant upon Graham for supplying "perfect" instruments, but he acknowledges that Graham assisted him in making observations and possibly in interpreting them. Although mathematical instruments usually measured existing phenomena, Bradley and Graham together used a zenith sector to reveal unexpected new truths in natural philosophy.[177]

3.2 The Shape of the Earth

Accurate instruments and their makers could also be crucial in settling long-standing scientific disputes. One of the most famous of these in the first half of the eighteenth century concerned the true shape of the earth: was it squashed at the poles (oblate) or squashed at the equator (prolate)? The story has been well told in many places, mainly from the perspective of the Parisian Academy of Sciences, which launched two dramatic voyages to Lapland and Peru in the 1730s and 1740s at the urging of Pierre de Maupertuis.[178] But the Royal Society also had a rather large stake in the whole question. Since Newton was no longer alive to promote his theory, that the earth was an oblate spheroid squashed by approximately 1 part in 230, the Society turned to George Graham to defend the Newtonian doctrine.[179] Graham made his own experiments in London, coordinated those of others in Jamaica, and supplied instruments for French measurements in Paris and Lapland. Despite these substantial contributions, Graham has not been particularly recognized by modern scholars for the major part he played in helping to solve the vexing question of the Earth's shape. Rob Iliffe takes more notice of Graham than most, pointing out that his instruments were used on the Lapland expedition because Graham was the "most prestigious instrument-maker of his day."[180] Graham's own measurements at his house on the Strand lacked the drama of those the French made in the Arctic wastes or the Andean highlands and if he has subsequently been overshadowed by Maupertuis's popular and engaging account of the Arctic expedition, it was not because of Maupertuis, who fully and effusively credits Graham in his publication.[181] Graham certainly lacked Maupertuis's dramatic flair, but he firmly grasped the mathematical principles at issue, conducted careful and observant experiments, and made the accurate instruments—a fixed-length pendulum clock, a theodolite, and a zenith sector—that measured the shape of the earth in tropical, temperate and arctic latitudes by two quite different methods.

The first of these methods depended on the comparison of the periods of pendulum clocks at different latitudes. Newton had noted in the *Principia* that several astronomers "have observed that their pendulum clocks went more slowly near the equator." Since the frequency of a pendulum is proportional to the square root of the gravitational force exerted on it, the clock runs slower when the force on it is less. What caused this lessening of force? Newton argued that the spinning earth tended, at the equator, to throw the pendulum off the earth's surface thereby reducing the force acting on it. Furthermore, the spinning earth also caused the earth to bulge at the equator. Thus the pendulum would be further away from the center of the earth and hence suffer less gravitational force. Newton thought that these two phenomena, acting in concert, comprised the main explanation for the clock running slower at the equator. That the pendulum lengthened as it got hotter near the equator, hence beating time more slowly, was only a minor contributor to the effect.[182] Graham hewed to this method, but he also wished to refine it further, taking into account the variation of the pendulum's beat due to temperature changes. So, in 1734, he set about seizing the opportunity that was "now offered of trying with the utmost Exactness, what is the true Difference between the Lengths of Isochronal Pendulums at *London* and *Jamaica*," by making the appropriate pendulum clock and the accompanying experimental protocols.[183] Graham gave "very full Directions" to Mr. Campbell, to whom the clock was sent in Jamaica. Graham gave no hints to Campbell as to what rate the clock ran at in London or how temperature variations influenced it, so that the "Experiment might be made with all possible Care and Caution, and without any Byas, or Prejudice, in favour of any Hypothesis."[184]

In fact, Graham had already established in London that his clock went 1 second a day slower for every 2 divisions of temperature rise on a specifically selected thermometer. Campbell's observations in Jamaica implied that any slowing in the pendulum's rate, in excess of the 9 seconds a day caused by the heat increasing the pendulum's length, would be due to the decreased forces acting on it at Jamaica. Thus, when Campbell found that the clock ran 2 minutes and 6 seconds a day slower at Jamaica, Bradley was able to calculate using Newton's "sine-squared" rule that the "Aequatorial Diameter is to the Polar, as 190 to 189," which he noted was "somewhat greater than what *Sir Isaac Newton* had computed from his Theory."[185]

There was another method to account for temperature change, namely eliminating it, either by building a clock that was not affected by temperature change, or by warming the clock up in

London so it was at the same temperature as the clock near the equator. Maupertuis adopted the converse of this latter method (heating the clock in the Arctic so it was at the same temperature as in Paris). Graham had already developed a version of a temperature-invariant pendulum in the 1720s but its design, with a glass column filled with mercury which expanded or shrank to maintain a constant center of gravity, was not robust enough for general use, let alone being taken to the tropics or the Arctic.[186] Given the practical and philosophical difficulties surrounding the measurement of temperature throughout the entire eighteenth century, Maupertuis's removal of temperature as a variable had much to recommend it.[187]

While he may have adopted a slightly different protocol from Graham, Maupertuis still needed his instruments for the expedition to Lapland, so he sent one member of the party, the Swedish astronomer Anders Celsius, to London in 1735. While in London, Celsius was elected a Fellow of the Royal Society, observed a lunar eclipse with Graham at his shop and residence in the Strand, and acquired from Graham not only a new sector similar to Bradley's, but also a theodolite, a clock, a pendulum, and a "machine" to measure the length of the pendulum, all of which were made by Graham.[188] Using Newton's sine-squared theory, together with the measurements from Graham's pendulum clock, Maupertuis (as had Bradley) calculated that the earth's diameter was longer by 1 part in 190 at the equator, and not, as Newton had calculated, by 1 part in 230.[189]

Comparing the period, or beating, of pendulum clocks was not the only way to measure the shape of the Earth; the other was to use a theodolite and surveyor's chains or rods to measure out a fairly large distance (60 to 70 miles) and then, using a zenith sector, to measure how much of a degree of arc that distance covered.[190] This method was physically more difficult and expensive, but it had the virtue of offering a direct measurement of the earth's surface; no complicated Newtonian assumptions about the net strength of gravitational attraction stood between the observations and the inferences from those measurements. Again, Maupertuis turned to Graham to make both the theodolite and the zenith sector for the requisite measurements. Perhaps hoping that it would reveal something as astonishing as it had to James Bradley in the previous decade, Maupertuis expressed every confidence in the zenith sector. It was, he enthused,

> ...of about 9 foot Radius. ...It was made at *London* under that ingenious Artist Mr. Graham, a Fellow of the *Royal Society*, who had exerted himself to give it all the Advantages and the all the Perfection that could be wished for. He had even taken the trouble to divide its Limb with his own hands. It were endless to give a particular description of everything that is remarkable in this Instrument.[191]

Graham's great reputation with Maupertuis and his colleagues came from three sources. Firstly, they wished to make arguments depending on the highest possible accuracy of theory, calculation, instrumentation, and measurement. Therefore, emphasizing the excellence of their instruments was a rhetorical necessity. This is not to say that the instruments may not have been as accurate as they claimed, but that without accurate instruments the measurements were severely devalued. Arguments took place in Paris after Maupertuis's return over the actual accuracy of the sector.[192] Secondly, as I have already shown, Graham was by then well known for the crucial role he had played in the discovery of the aberration of starlight. And thirdly, Maupertuis and his colleagues were hoping to prove the Englishman, Newton, right, and Jacques Cassini (Cassini II), their compatriot, wrong. Graham was the finest English instrument maker of his generation, a member of the same

Royal Society that had become synonymous with Newton, and the equipper of the Royal Greenwich Observatory run by Halley and supervised by the Royal Society. In a very literal sense, Graham's instruments were Newtonian or, what amounted to the same thing, they were English astronomy manifested; the prestige of Newton and the excellence of the English instruments were inextricably tied together.

Maupertuis's expedition had the full support of the Royal Society. Maupertuis was already a Fellow and the other four main members—Anders Celsius, Alexis-Claude Clairaut, Pierre-Charles Le Monnier, and Charles-Etienne Camus—were elected to the Society as a consequence of the expedition.[193] It was hoped that Maupertuis's measurements near the polar circle would confirm the series of experiments that had already been undertaken by three of the Society's members—Graham, Campbell, and Bradley—in London and Jamaica using Graham's pendulum clock. A steady stream of letters arrived from France and Sweden to inform the Royal Society of the progress of the Lapland expedition.[194] In one of these letters Maupertuis requested advice from Bradley about the observational corrections that the aberration of starlight demanded. Of Graham's sector he wrote: "notwithstanding the high idea we had of Mr. Graham's ability, we could not but with astonishment see... [that] the arc of the limb did not differ but one second from what it ought."[195] Graham himself could scarcely wait to hear the final results of the expedition. Although Le Monnier had to put him off, telling him to wait until the book was published in a couple of months, Graham was duly given one of the seventeen copies of *La Figure de la Terre* that Maupertuis sent to the Royal Society.[196]

In the book, which was immediately translated into English, Maupertuis told how his team had fixed Graham's sector in the plane of the meridian to measure the angle at which a particular star near the zenith passed the meridian (the star's altitude); they then moved the sector south to another mountaintop and repeated the same measurement. The difference between the two altitudes was due to their having moved south on the globe. The academicians then established the distance between the two mountaintops by surveying a series of triangles, with Graham's theodolite, and measuring several of the baselines with wooden poles. Having measured the distance between the two points and the difference in angles, they calculated that the length of one degree of arc near the polar circle was longer than at Paris; therefore they concluded that the earth was squashed at the poles. This seemed, at first glance, to vindicate Newton. At the very least, Cassini and his fellow Cartesians were wrong in saying that the Earth bulged at the Poles. Of the disagreement between Newton's prediction of a bulging at the equator of 1 part in 230 and Maupertuis's own measurement by pendulum clock and zenith sector of 1 part in 190, Maupertuis had less to say. When he could avoid it no longer, he gave his preference to observation over theory; his measurements with the sector, he argued, were more precise than Newton's theory, since his angles were accurate to within a second and his lengths were accurate to within inches.[197]

Thus, by 1738, Graham had helped others measure the shape of the earth with his pendulum clock and had seen his instruments triumph in the skilled hands of Maupertuis in Paris and Lapland, Campbell in Jamaica, and Bradley in London. The question of the shape of the earth was settled and Graham himself could say with satisfaction to Bradley that "Sir Isaac Newton would have been much pleased... had he been living."[198]

3.3 The Search for the Great Southern Continent

Whereas the shape of the earth could be predicted by the mathematical principles of natural philosophy and confirmed with the instruments of mixed mathematics, the outlines of the land masses on that earth had no *a priori* shape. Indeed, any attempts at theoretical cartography were justly ridiculed in the eighteenth century. Navigators in Britain derided theoretical geographers who predicted the existence of a great southern continent and Louis de Bougainville, who had sailed to Tahiti in the 1760s, proudly called geography a "science of facts" and heaped scorn on the "class of indolent writers, who in their closets reason *in infinitum* on the world."[199] Bougainville's venom was matched by that of Jean-François de la Pérouse, who sailed into the Pacific in the 1780s, and spoke just as contemptuously and scatologically of "the system makers, who sit down in their closets, and there draw the figures of lands and islands."[200] To settle the most basic of geographic facts, the Royal Society had to turn to the Admiralty, the only corporate body in Britain which could bear the cost of that crucial instrument of geographic discovery, the ship, which in the hands of the naval officer, James Cook, contributed greatly to the mixed mathematical sciences of geography and navigation in the 1760s and 1770s.[201]

The ostensible mission of Cook and his crew on his first voyage was to observe the transit of Venus and hence to establish the absolute size of the solar system.[202] The actual one, however, was to map and lay claim to a possible new empire in the South Pacific.[203] To accomplish his mission, Cook took along telescopes, lenses mounted on transit instruments, sextants, barometers, and watches made by the Fellows Short, Dollond, Ramsden, and Graham, as well as clocks, quadrants, thermometers, and transit instruments by John Shelton and John Bird.[204] The navigational instruments that he took on board worked in concert with the mathematical techniques of astronomers. In particular, the lunar method for finding longitude at sea finally became reliable on Cook's first voyage. Although taking lunar distances was complicated, navigators generally preferred the superior accuracy of this new method, which used the moon as a celestial clock that could be "read" afresh each day, even though it required up to four observers, several instruments, various mathematical tables, and numerous calculations.[205] More accurate still were the methods that could only practicably fix longitude on land—namely, measurements of eclipses and transits. Thus, when he could, Cook allowed his astronomer, Charles Green, to set up temporary observatories on land. When subsequently compared with those that had been made or calculated in London, Green's measurements would fix longitude at certain places with greater accuracy.

While the track of the ship at sea could be relatively easily projected onto a map gridded with latitude and longitude, making a map of coastal outlines from the ship was more complex. Cook used his knowledge of the ship's position to infer the outline of the coast by a running survey. Keeping the boat as stationary as possible, he would take the bearing of a prominent coastal feature. He would then move to another position and take a new bearing of the same feature. He calculated the interval between the two positions either by using a log to measure the distance travelled or by determining the latitude and longitude of each position. Once the baseline was established and the angles were measured, the latitude and longitude of the coastal feature could then be calculated.[206] With such a method he could construct a map by using the ship to probe and skirt around, yet never touch, the coastline. Thus the methods that allowed Cook to "steer a course also enabled him to leave the coastlines he sighted where they were. This was the essence of the maps he made, that they did not mirror the appearance of natural objects, but preserved the trace of encountering them."[207]

When Cook encountered New Zealand, he was eager both to map its coastline accurately and to prove that it did not belong to the supposed Great Southern Continent. He was often torn, in other words, between lingering to draw convincing coastal outlines and forging ahead to close the curve that he was drawing around the land.[208] While he is justly renowned for the unprecedented accuracy of his maps, the errors he made reveal not only the subtle nature of the instrument he was using to augment the science of geography—namely, his ship—but also the faults generated by his own impatience to "discover" a new island.[209]

Figure 2 both outlines and re-enacts Cook's encounter with New Zealand in 1769-70.[210] While the tracks on the map provide a record of "where Cook went" and an indication of the systematic nature of his geographical investigations, they also trace his methods of constructing the map itself. His map of New Zealand thus does two things at once. Firstly, it disconnects New Zealand from the *Terra Australis Incognita* by demonstrating, as Cook noted in his own journal,

> ... that this country which before now was thought to be a part of the imaginary southern continent, consists of two large islands divided from each other by a strait or passage 4 or 5 leagues broad. They are situated between the latitudes of 34°and 48°S and between the longitudes of 181°and 194°W of Greenwich.[211]

Secondly, it establishes the one-dimensional outline of that country as constructed from the running survey conducted from the *Endeavour*.

The zigzagging of the track of the ship, as opposed to the outlines of the land, inscribes the daily vagaries of the weather, coastal landmarks, the need to provision the ship and the crew, and inter-

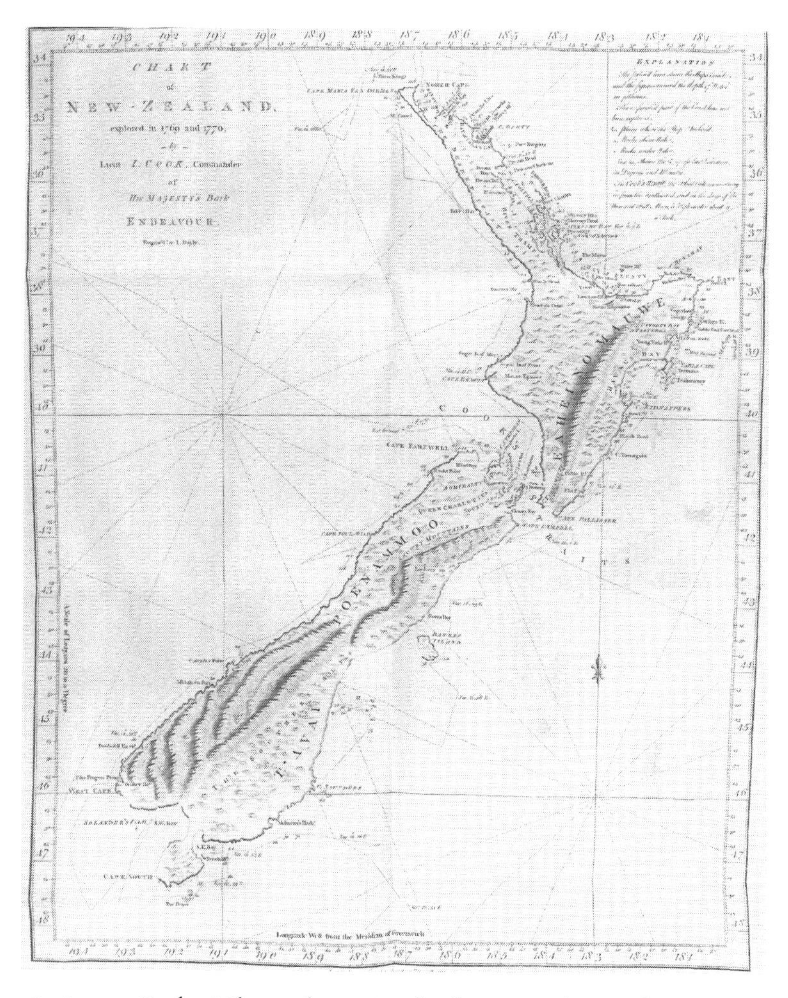

Figure 2: James Cook, "Chart of New Zealand," in Hawkesworth 1773; image by permission of the Lilly Library, Indiana University, Bloomington, Indiana.

actions with the local Maori.[212] For most of this business, no great accuracy was required. To bring a new country convincingly into being, a surveyor had to de-emphasize the ship's somewhat random track and thicken up the coastline. Cook accomplished this "by some hundreds of observations of the sun and moon and one transit of Mercury made by Mr. Green who was sent out by the Royal Society to observe the Transit of Venus."[213] His running survey became less accurate, however, whenever the ship had to retreat from the coast, since it became more difficult to judge how close the prominent landmarks selected for the running survey were to the coastline itself. Ideally, of course, the landmarks would actually be part of the coastline, but in some cases they lay somewhat inland. From latitude 42S to 45S along the eastern coast of the Southern Island of New Zealand (Te Wai Pounamu), for example, the coastline was low lying and backed by substantial mountains. Cook saw the mountains but was too far distant to see the coastal plains, or perhaps he misjudged their extent. Cook's editor and biographer, John Beaglehole, notes that if one looks at a modern map of New Zealand in this region and follows the contour lines at 1000 feet, "one will have the line of Cook's chart on this eastern coast."[214] Cook's map of the land leaves vestiges of his own uncertainty; the coastline is at times thinner here to match the greater distance of the ship's track from the land.

On his second voyage, Cook set out not to discover new lands but rather to prove that the great southern continent was an illusion of two millennia. His zigzag path encircling the bottom of the world, which he proposed to the Admiralty in 1772, was carefully planned rather than haphazard. Cook proposed to circumnavigate the globe in its high southern latitudes near the South Pole, taking a generally "Easterly Course on account of the prevailing westerly winds," making sure the ship was not at sea during the southern winter, and using New Zealand as a base for resupplying the ship's water. If he could keep to this course, zigzagging to cover a reasonably wide band of latitude, and if he saw no land, then he could assume that no continent lay in these immediate latitudes; at worst he had missed some small islands.[215] He was very careful to maintain that his probing had not ruled out the existence of a considerably shrunken continent in even more southerly latitudes. Indeed, Cook called the existence of such a continent "more than probable." Evidence for this unseen continent included the "excessive cold" and the existence of "vast floats of ice," which Cook thought were first made on land.[216] When faced with intense cold and the imminent danger of being sunk by floating ice-mountains, he was unable to traverse the latitudes closest to the pole where this land might lie. In general, however, Cook stuck to his plan (Figure 3).[217] He was proud of his ship's performance, writing at the end of his account of his second voyage:

> I have now made the circuit of the Southern Ocean in a high Latitude and traversed it in such a manner as to leave not the least room for the Possibility of there being a continent, unless near the Pole and out of reach of Navigation. ... Thus I flatter myself that the intention of the Voyage has in every respect been fully answered... a final end put to the searching after a Southern Continent, which has at times engrossed the attention of some of the Maritime Powers for near two Centuries Past and the Geographers of all Ages.[218]

The logical uncertainties of Cook's enterprise—to prove that something did not exist—are nicely summed up in the title of one book that described the voyage, J. Marra's 1775 *Journal of the Resolution's voyage: By which the non-existence of an undiscovered continent between the Equator and the 50th degree of southern latitude is demonstrably proved.* To demonstrably prove the non-existence of the undiscovered is not so easy a task as it might sound. The tracks that Cook made in the high latitudes of the Southern Hemisphere (see Figure 3) map nothing but themselves. They bear a relationship not to existing land masses but merely to geometrical points in space, the lattice of latitude and longitude. Cook's discoveries (and non-discoveries) in the South Pacific would have

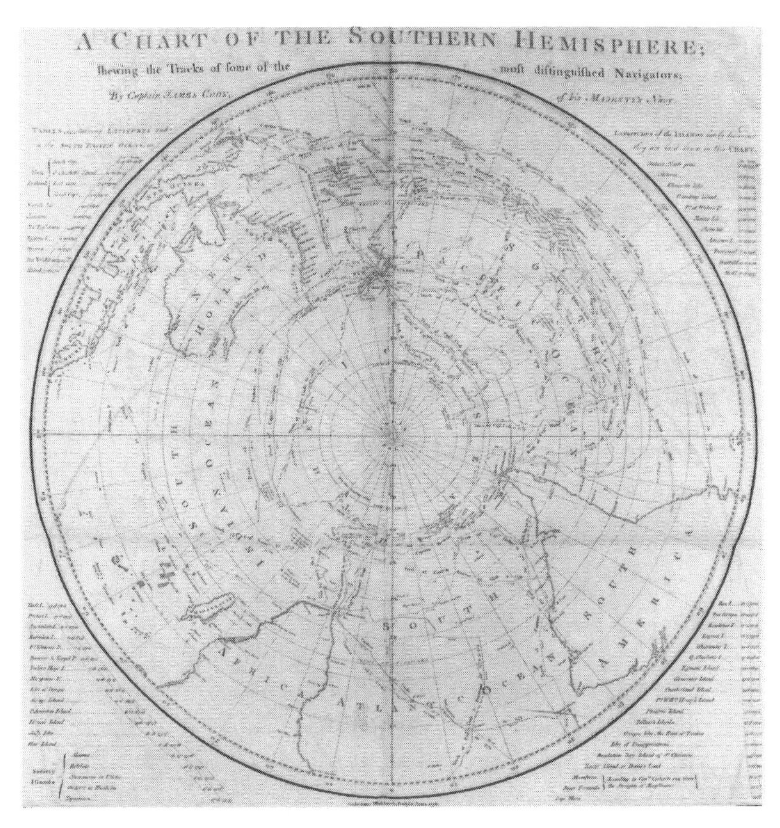

Figure 3: James Cook, "A chart of the Southern Hemisphere," facing p. 1, in Cook 1777; image by permission of the Lilly Library, Indiana University, Bloomington, Indiana.

been impossible without the many scientific instruments he carried on board his ship, which allowed him to navigate and map with a high degree of precision. Cook himself, whose ships were outfitted with mathematical and optical instruments on the advice of the Royal Society, practiced the kind of meticulous, methodical science that the Society most highly valued, using mixed mathematical measurement to make new discoveries and settle long-standing disputes about the world. But he also gave credit to the men who made such reliable measurements possible, lauding mathematical instrument makers for "the improvements and accuracy with which they make their Instruments."[219]

3.4 Magnetic Variation

Instrument makers at the Royal Society worked constantly to improve the accuracy of their instruments. Occasionally, when an instrument did not work as well as it should, they turned to experiment to establish whether and to what extent unknown phenomena were affecting its performance. In so doing, they demonstrated full control of Royal Society protocols for scientific work. Whether or not the instrument makers were able to explain the new phenomena they revealed, their experiments were important in advancing natural philosophical knowledge and debate.

The magnetic compass, for example, was a useful but problematic navigational instrument. By the time Graham came to look at it more closely in the 1720s, three of its defects were already well known. Firstly, at a given location the compass did not point to the fixed point of the north pole (geographic north) but to another fixed point (magnetic north). The angle between these two points was called the magnetic variation. Secondly, that variation itself varied as the compass was moved from place to place. Thirdly, the magnetic variation at a given place changed over time. This last variation is also known as the secular magnetic variation. These three defects were by no means fatal to the utility of the compass; once understood, they could be measured and taken into account.[220] But Graham wondered whether there were further defects that had not yet been identified. Did the magnetic variation at a given place vary also from day to day?

He worked at his own behest and received sole credit for his demonstration that the needle's variation was not stable, but rather underwent a diurnal change. Graham made no voyages nor proposed any theories. He stayed at home and, with his own instruments, made a "fundamental discovery."[221] The success of his experiments depended upon accurate and regular observations of an already well known phenomenon and his ability to design instruments and experimental protocols that were sensitive and exact. Graham's compass needles were delicately made and carefully balanced on a pin "of steel hardned, and ground to a fine Point." He removed all iron and steel from the room, and carefully fixed a meridian line of fifteen feet long that ran above the compass. Below the compass itself, he made a brass plate representing minutes of arc that he divided into 10' sections. Above the compass box he placed a convex glass that, acting as a magnifying lens, enabled him to estimate (within 2') the position of the needle. Before making "any Trials," he repeatedly moved the needles and noted that they always returned to their initial position. With two compasses having needles of different weight, Graham settled down to over a year's work, making "above a thousand Observations in the same Place."

Such meticulous observation had its rewards and Graham's instruments and methods revealed a diurnal variation of 35' or so a day, an amount that was both unexpected and philosophically intriguing, though of little practical import.[222] The variation was never exactly the same from one day to another: "the only thing that has any appearance of Regularity, is that the Variation has been generally greatest... between the Hours of Twelve and Four in the Afternoon, and the least about six or seven in the Evening."

As to "what the Cause is," Graham concluded, "I cannot say."[223] John Canton, a contemporary electrician at the Society, attempted speculations related to the sun heating the region to the east of the compass in the morning and the west in the evening, but Graham demurred.[224] Nonetheless, he was certain of his facts, secure that the changes were not generated by the vagaries of the instrument. Graham's experiments did not help him or anyone else make better magnetic compasses, but his discovery of diurnal magnetic variation did lead to a better and more precise understanding of their flaws and demonstrated that hitherto unknown natural phenomenon could be revealed by patient work with the best available instruments.

3.5 The Dispersion of Light

Whereas Graham's delicate investigations of the magnetic variation had no immediate effect on the construction of practical compasses, John Dollond's experiments on the dispersion of light had an immediate and major impact on the making of refracting telescopes which, until the 1750s, suffered from two major defects. The first—the problem of spherical aberration—was that any lens ground to a spherical shape could not focus perfectly. Its correction was a technical, not a theoretical, challenge. The correct parabolic or hyperbolic curve of the lens was known, but putting theory into practice was exceptionally difficult.[225] The second—the problem of chromatic aberration—was that a simple convex refracting lens would bend light of one color to a different focal point from that of another color; hence any image formed from transmitted white light would be fringed with color. According to Newton's Experiment VIII (in Part 2 of Book I of his *Opticks*) upon which this assertion was based, this defect could not be solved. Any combination of refracting lenses would necessarily leave colored fringes around images. The gravity of the situation led him to declare that "the Improvement of Telescopes of given lengths by Refractions is desperate." This was Newton's final public claim. Earlier in his life, he had thought an achromatic lens might be possible, but had never been able to make one.[226]

So great was Newton's authority that no one repeated Experiment VIII for more than fifty years. Indeed the first challenge to its authenticity was not simply to repeat it, but rather to argue that it must have been interpreted incorrectly. In 1747 Leonhard Euler plucked up enough courage to assert that Newton must be wrong since there is one supreme example of a refracting lens that gives images free of any chromatic aberration: the eye itself. After all, we do not see colored fringes around objects when we look at them, nor would God design an imperfect lens. Euler therefore recommended a design mimicking the eye.[227] We will hear of this analogy again, so it is important to note that it is flawed. The compound lens system of the eye (a convex lens surrounded by convex transparent humors) delivers chromatically aberrant light to the retina but the brain removes this defect so we "see" clearly. A lens, designed to copy the eye, cannot work as promised.[228] Nonetheless, the analogy motivated Euler to think, in opposition to Newton in his *Opticks*, that chromatic aberration could be removed.

Such a challenge to England's greatest natural philosopher, arriving in 1749 due to the very slow publication of the *Mémoires* of the Academy of Sciences in Paris, could scarcely go unanswered by the Society that had been adorned by Newton's presidency. But the Royal Society of London had no mathematicians or theoretical opticians of Euler's eminence to rely upon; nor did its most famous practical optician, James Short, himself answer Euler. Perhaps he had no interest in improving the technology of refracting telescopes since he dominated the trade in reflecting telescopes, whose one great advantage was that they did not suffer from chromatic aberration. Instead, sensing an opportunity to introduce himself to the world of letters, another paladin appeared before the Society: a former silk weaver and self-taught London optician, John Dollond, who was not yet a Fellow of the Society, but was vouched for by Short.[229]

Figure 4: John Dollond, painted by William F. Witherington, after the painting by Benjamin Wilson, image by permission of the Royal Society of London

How did Dollond have the knowledge or confidence to reply to Euler? Dollond's education remains very obscure. He knew French; possibly because of his Huguenot background. His portrait shows an optical treatise prominently on the table in front of him. (Figure 4) This treatise, judging from its appearance and size, is most likely Robert Smith's *Opticks* (written for autodidacts) or Newton's work itself, both of which Dollond surely studied. J. Kelly, his grandson-in-law, notes rather vaguely that while he was still a silk weaver Dollond's "leisure hours were engaged in mathematical pursuits" which led to a "considerable proficiency in optics and astronomy... [and] a perfect knowledge of algebra and geometry."[230] Kelly's family-commissioned biography does not tell us how he attained these proficiencies. It does however demonstrate that an ambitious and clever silk weaver in Spitalfields could educate himself to a remarkable degree in mid-century London, whether through reading, attending popular scientific lectures, or belonging to mathematical clubs. While there is no specific trace of John Dollond in the records of the Spitalfields Mathematical Society, the rather remarkable eighteenth-century-London club for artisans and merchants that Stewart and Weindling have so ably brought to life, it seems highly likely he had some contact with it, given the Society's strong Huguenot associations and its location in Spitalfields.[231]

Like Euler, Dollond still did not repeat Newton's Experiment VIII. Instead, in the response to Euler that he wrote on behalf of the Royal Society in 1753, Dollond contented himself with reasserting Newton's authority, noting that

> ...the famous experiments of the prism... sufficiently convinced that great man, that the perfection of [refracting] telescopes was impeded by the different refrangibility of the rays of light... [it is] therefore somewhat strange that anybody now-a-days should attempt to do

> that, which so long ago has been demonstrated impossible.

Dollond went on to argue that Euler's error arose from his wanton adoption of "a certain hypothesis... destitute of support either from reason or experiment."[232]

This may seem a rather extraordinary way to address Europe's greatest pure mathematician. However, it was not out of place at a Society that gloried in a manly contempt for speculation and saw no need to repeat experiments that had already been added to the stock of knowledge and certified as true by no less than England's greatest scientific authority, Sir Isaac Newton. Dollond was not merely an optical autodidact. A year earlier he had gone into partnership with his son Peter, who had been running an optician's business since 1750. Dollond & Son made and sold a wide range of wares at the Golden Spectacles and Sea Quadrant on the Strand in London.[233] So when John Dollond flatly stated that it was "impossible" to make a refracting lens that corrected chromatic aberration this was more than his own opinion; it must also have reflected the common knowledge of the entire London optical community.

In the summer of 1755, the same year he became a member of the Spectaclemakers' Company, Dollond received a letter from a Swedish astronomer and mathematician, Samuel Klingenstierna, who had read the exchange between Euler and Dollond. Klingenstierna stated, as had Newton and Dollond before him, that "the Aberrations, arising from the different refrangibility of the rays of light cannot be corrected by any refractions whatever." However, crucially, he argued that Newton's general law of dispersion—that the width of a spectrum depended only on the angle of refraction of the mean ray and not the material nature of the prism—was wrong, and he concluded that there ought to be an "examination of this matter by further and more certain experiments."[234]

This letter finally prodded John Dollond into action and, in early 1757, he repeated Experiment VIII and made the appalling discovery that Newton had erred.[235] Dollond found that there were two different kinds of glass—crown glass, and lead or English flint glass—with similar refractive powers. The refractive power of glass is its ability to bend the mean ray (also known as the colored ray of light from the middle of the spectrum). However, these two kinds of glass had very different dispersive power, namely the ability to bend the colors at each edge of the spectrum and hence make a wide or narrow spectrum from white light. Therefore, it followed conversely that for the same dispersive power the refractive power of one substance was greater than the other. In this case, Dollond could arrange two prisms so that the dispersive powers pretty much cancelled each other out but there still remained a net refraction (Figure 5). Thus the incoming white light was bent by the compound prism but not broken up into all the colors of the spectrum: the transmitted light was much freer of color.

By showing that white light could be refracted but not necessarily dispersed into all the colors of the spectrum, Dollond was the first to demonstrate with substantial experimental proof that dispersion was not dependent upon the refraction of the mean ray alone, or as he put it, "these experiments clearly prov'd that different substances diverged the light very differently" for a very similar degree of refraction.

In the 1758 *Philosophical Transactions* paper in which he presented the results of his experiment, Dollond suggested that it followed from the arrangement of the prisms that a thinner, less refractive flint concave lens combined with a thicker, more refractive crown convex lens could bend light to a focus with very little net dispersion. Completely absent, however, are any further details of the lenses' construction. Dollond only hinted at how he had solved the "very considerable" difficulties of removing the spherical aberra-

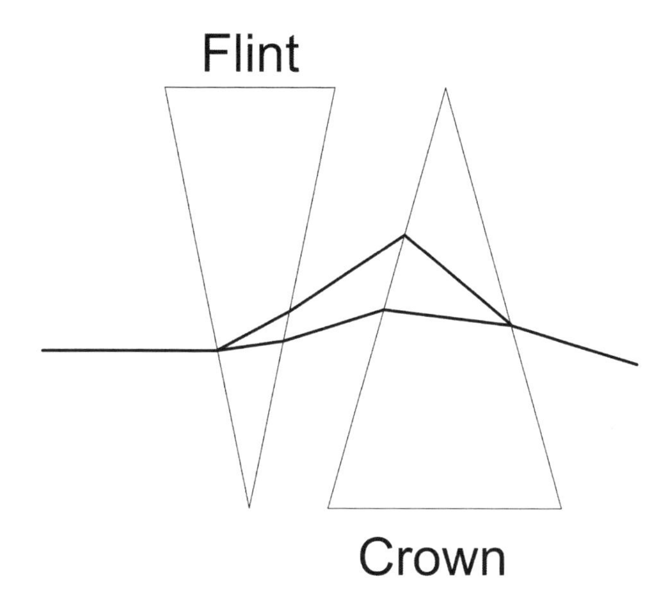

Figure 5: Dollond's arrangement of prisms

tion. He had noticed that he could match the spherical aberration of the crown to the flint; since "the refraction of the two glasses were contrary to each other, their aberrations... would intirely vanish." But all four surfaces had "to be wrought perfectly spherical" a process that required the "greatest accuracy." The resultant telescope was, James Short averred in an accompanying note, excellent.[236]

It is important to note that the artifact and the optical principle worked in tandem for Dollond. An interested reader who had not yet looked through one of the wonderful new devices could persuade himself of the telescope's excellence by accepting the truth of Dollond's experiments and the conclusions he drew from them; the achromaticity of the lens followed from the achromaticity of the arranged prisms in the experiments. Conversely, if the reader doubted the experiments he could look down a Dollond refractor himself and experience white light refracted but not dispersed into a full spectrum—something that Newton, in the *Opticks*, had said was not possible.

The paper and the artifact caused a sensation throughout the European optical community. Astronomers travelled to London to buy the new telescopes, kings and princes collected them, practical opticians tried to copy them, and theoreticians set about mathematizing the surprising result. For example, the Swedish astronomer, Bengt Ferrner, purchased two telescopes in 1760 on a visit to London. The first, which cost £7.7s, was for the Royal Swedish Academy of Sciences, and the second for Klingenstierna, who urgently wanted to use, inspect, and have copied the lens he had helped bring into being. Subsequently the Academy acquired two other Dollond achromatics, one of which had been the personal property of the Swedish King. The Duke of Orleans had a Dollond achromatic telescope valued at 1,400 *livres* (£70) in his collection.[237]

Dollond was immediately awarded the Royal Society's highest honor, the Copley Medal, for his "long course of Experiments which he contrived with Skill and Executed with the greatest Accuracy and Exactness." In granting the award to Dollond the Society had, Lord Macclesfield noted, "thereby given a convincing proof of their impartial regard to merit wherever they find it, without confining the favours of the Royal Society solely to the Members thereof."[238] Three years later the Society expanded its favors to include the shopkeeper, and Dollond became a Fellow in the last year of his life.

Chapter 4

Rational Instrument Design

In their continual quest to design better and more accurate instruments, instrument makers did not confine themselves to investigating the natural world alone. They also sought a more perfect understanding of the world of artifacts: that is, their own instruments. To make quadrants, sextants, or sectors that had almost perfectly divided angles, eighteenth-century instrument makers put into practice geometrical constructions through their skilful use of compasses and straight-edge rulers. And when the materials from which an instrument was made affected its performance, as was the case with a clock pendulum or a telescope lens, they designed new instruments—the pyrometer and the vitrometer—to measure those materials' properties.[239] Thus, by applying geometrical reasoning when appropriate and undertaking empiricist experiment when necessary, they achieved a rational instrument design. Their improved instruments made possible better measurements and their improved measurements made possible better instruments thereby creating a virtuous circle, very much in keeping with the Royal Society's emphasis on improving knowledge through diligent observation and experiment.

4.1 Rulers and Compasses

Mathematical instrument makers made simple geometrical theorems real. They used Euclidean straight-edge and compass constructions to carry out "three basic operations: (i) drawing a line segment connecting any two points... (ii) extending a line segment by any given line segment, (iii) drawing a circle with any given point as center and any given line segment as radius."[240] Yet these constructions are themselves mechanical, as Newton pointed out in his Preface to the *Principia*: "the description of straight lines and circles, which is the foundation of *geometry*, appertains to *mechanics*." The better the instrument maker, the straighter the lines and the more perfect the circles; and "if anyone could work with the greatest exactness, he would be the most perfect mechanic of all." Only when the beginner has mastered the art of precision, Newton concludes, may he "approach the threshold of *geometry*," and with the tools of geometry "reduce the art of measuring to exact propositions and demonstrations."[241]

Eighteenth-century instrument makers laid down straight lines and circles on metal with a physical action yet divided those lines and circles by putting into practice simple Euclidean propositions with their compasses and rulers. Their work thus required both exact mechanical execution and an intellectual grasp of mathematical theory. For the great majority of Fellows at the Royal Society, as we have seen in Chapter Two, mixed mathematics of this kind was superior to pure mathematics, which they regarded as being completely removed from its origins and usefulness in everyday life. Such bastardization horrified some continental philosophers, like Jean LeRond d'Alembert, who thought that "pure mathematics should be independent of and prior to mathematical physics." For others however, such as Immanuel Kant, mixing up the physical and the mathematical was "not a defect but a virtue."[242]

George Graham demonstrated the virtues of mixing mechanical brilliance with a sound understanding of geometry when he made fine divisions of angles down to the last second. As we have already seen, his zenith sector, with its precisely and minutely divided angles, was crucial to several important mixed mathematical investigations in the second quarter of the eighteenth century. Another device for measuring angles, the large quadrant that he made for the Royal Greenwich Observatory, cemented his reputation. Contemporary accounts of that quadrant give us some clues as to Graham's methods. In 1725 Halley was granted £500 for new instruments, including Graham's quadrant, and a Royal Society committee reported on the work. One of the committee members was George Graham himself, who wrote that Halley had indeed procured the quadrant as required by the committee and that it had "been made and fixd with the greatest exactness in every particular" and at the best possible price consistent with "putting the Instruments in such a state of perfection."[243]

In an age of much corruption, it seems that "Honest" George Graham lived up to his name, for the Observatory not only got its quadrant, it got a quadrant that was indeed of unparalleled accuracy. The praise for this instrument has been consistent ever since it was made. In 1738, Robert Smith, Master of Mechanics to his Majesty, asserted that its "particular accuracy... excels all others... owing to the extraordinary skill and contrivance of Mr. *George Graham*." Nearly fifty years later, Nevil Maskelyne, the Astronomer Royal, called it "excellent... [of] a degree of accuracy unknown before," while the Cambridge fellow William Ludlam thought that Graham "carried the art of constructing and graduating [this] instrument... to such perfection, that from this time we may date a new era in Astronomy." English-made copies of Graham's instrument went on to fill the observatories of Europe.[244] Graham did not manufacture the whole quadrant himself; Jonathon Sisson undertook the construction of the brass arc and the iron frame. Robert Smith, in his account of the quadrant, is careful to state the difference between the "contriver" of the instrument, George Graham, who directed the "whole design" and "was pleased to perform the divisions of the arch and all the nicer parts of the work with his own hands;" and the "inferior workmen" whom he supervised.[245]

As Allan Chapman has demonstrated, Graham's importance lay, not only in the delicacy with which he divided the arc of the quadrant using standard geometric techniques, but also in his innovative use of the vernier scale to read the finished divided arc, his use of a scribing compass, instead of the more clumsy knife and ruler to mark the divisions, his introduction of a 96-part scale to cross check the 90-degree scale, and the massiveness of the superstructure which supported the arc of the quadrant. All of these techniques, which were copied and improved upon by the generation that followed Graham, helped to establish the English instrument-making trade as preeminent in Europe. Graham's commitment to making geometry real is evident in his techniques. When possible he chose bisection, which could be achieved using straight-edge and compass geometry; only when absolutely necessary did he trisect or quinquesect. For example, he chose a 96-part scale to cross-check the 90-degree scale, because the former could be continually bisected while the latter could not. Graham's bisection technique was as follows:

> ...from the [zero] point the radius compass could strike off 60 degrees, on the [Euclidean] principle that a radius, as a chord, will divide into a full circle six times exactly. The 60 degree space was accurately bisected with a shorter compass. To do this, two bisecting arcs were drawn from each end of the 60 degree arc, in such a way that the scratches they made on the metal just failed to

> touch. ... Graham found it easy to fix the precise point between them by the aid of a strong magnifier. ... Once the 30 degree point had been found, the compass was set into the 60 degree mark, and a full 90 degree arc constructed. Each 30 degree space was likewise bisected to contain 15 degrees.[246]

Graham could achieve bisection of a 90-degree scale only down to 15 degrees, after which he had to trisect or quinquesect: operations which, since they relied on successive approximations and comparisons, were not possible in plane geometry.[247] Once Graham had reached the 15-degree arc by bisections, he then copied that arc on to another piece of brass. To trisect he used another compass which he set to an estimated one-third division of the 15-degree arc. From each end of the arc he inscribed a mark on the circumference. He then measured the two interior marks using the compass. If, for example, the initial one-third division was too large, the two interior marks would be too close together when compared with the compass. Thus the compass would be closed slightly, and the process repeated until the two interior marks were the same distance apart as the measuring compass. Once that occurred, the compass would be used to make the 5-degree divisions on the original arc.[248]

The process of quinquesection used an identical approximation procedure but was even more painstaking. These methods were not only tedious; they were not geometrically pure, having been forced upon Graham by the historical accident of the 90-degree right angle which was too entrenched to be replaced by a 96-degree one. Because of the difficulty and precision of the task, it was very important that Graham himself laid hands on the instrument as he applied geometry to mark off angular divisions. His skill, workmanship, and personal integrity, as much as his grasp of geometrical principles, ensured the accuracy of the instrument that bore his name.

4.2 Machine Tools

No machine, only the simple tools of geometry, intervened between Graham and the angular divisions on his quadrant. But makers such as Graham were few and expensive, and the demand for accurately divided instruments intensified throughout the century. The British state, which had a major interest in the production of such instruments to assist naval and commercial interests, offered substantial rewards to anyone who could successfully and cheaply meet the demand. A meticulous and talented maker could accurately divide a few dozen instruments a year; a well-organized workshop of lesser men could less accurately divide hundreds or even thousands a year.

Only a machine could produce instruments that were both accurate and abundant. However, the instrument maker who solved the problem of designing and building such a machine would no longer be directly applying ruler and compass geometry to divide the instrument. He would never touch it. Instead the entity that produced the accurately divided angles, an endless screw thread, would lie buried deep within the machine itself. An instrument of very high accuracy could now be divided by relatively unskilled workers.

Jesse Ramsden was paid £615 by the Board of Longitude for his invention of such a machine. In return, Ramsden was to instruct other instrument makers in its use and construction, divide other makers' octants and sextants at a fixed rate, and write an account of the machine tool. Ramsden's "Description of an Engine for Dividing [Circular] Mathematical Instruments" appeared in 1777 and was followed in 1779 by another pamphlet that described a new machine for dividing straight lines.[249] Ramsden was ordered to charge three shillings for dividing octants and six shillings for dividing sextants on the former machine. The Board of Longitude printed 500 copies of the pamphlet describing its operation and construction.[250] The Royal Society also expressed its admiration for this machine tool, "by which an accuracy unheard of before except in the most expensive instruments had been transferred to the cheap sextants used by our Merchant Sailors," and gave Ramsden the Copley Medal in part for its construction.[251] But amid all this success for solving an urgent and practical problem lay a threat to Ramsden's status as a producer of scientific knowledge.

In the pamphlet of 1779, Ramsden boasted that "any line of equal parts... may be divided without an error of the 1/4000th part of an inch" and went on to note that "as this can be done by any indifferent person, and so very expeditiously, its uses for dividing all sorts of navigation scales, sectors, etc. must be obvious."[252] Ramsden's claim that "any indifferent person" could operate his engine accurately reflected the impact of surrounding industrial developments—the division of labor and use of machines operated by deskilled labor—on the instrument-making trade. It was a dangerous boast; now, accuracy was produced by the machine and not the maker, whose role had become that of a machine tool builder rather than a skilled artisan. [253] Ramsden's circular dividing engine, which is still in existence and is now at the Smithsonian Institution, is shown in Figures 7 and 8.[254] So important was this machine to Ramsden, that it appears in the foreground of his portrait by Robert Home which is now in the possession of the Royal Society (Figure 6). The large bell-metal wheel, mounted on a mahogany stand (Figure 7), was 45 inches in diameter, and its edges were cut into 2,160 teeth, on which an endless screw thread acted (shown from above in Figure 8). Six turns of the handle moved the wheel one degree, and the brass circle to be divided was fixed to the wheel. The divider (in Figure 8), which slides along the frame, was fixed when it was over the circle to be divided. The screw thread was rotated a small amount, corresponding to the fraction of a degree wanted for the instrument being divided, and the divider was pressed down and slid towards the center of the engine to make a mark on the instrument. Thus the accuracy of the screw thread was responsible for the accuracy of the division. Arithmetic had replaced geometry; rather than using rulers and compasses, the divider turned the screw thread the appropriate fraction and marked the instrument.

Since an accurate screw thread was so important to the operation of the dividing engine, Ramsden also made a screw thread cutting machine and was required by the Board of Longitude to publish details of its working.[255] The degree of accuracy achieved by Ramsden (more than $1/1000^{\text{th}}$ of an inch) for his instruments of brass was not equalled on iron- or steel-manufactured goods until the latter half of the nineteenth century. Yet Ramsden tells the reader nothing of how he made the reference or master screw thread against which copies were made. That secret he kept to himself, hidden from the prying eyes of visitors to his workshops or the demanding Commissioners of the Board of Longitude. Ramsden's secret has somewhat eluded even Allan Chapman who has skilfully replicated the methods of many of the instrument makers he has studied. Chapman notes that Ramsden concealed his methods of making the master screw thread and concludes that he probably produced it on his own specially adapted horological fusee lathe. Yet this machine also relied on a master screw thread that Ramsden made to a standard of accuracy that is still very difficult to account for.[256] While the dividing engines were meant to deskill those artisans who made ordinary instruments, the man who made the dividing engines—or more precisely the master or reference screw threads that lay at the heart of the production of accurate instruments—had to be supremely skillful. As the century drew to a close, however, the skill involved in machine tool manufacture became less and less a recommendation for membership in

Figure 6: Jesse Ramsden, painted by Robert Home; image by permission of the Royal Society of London.

the Royal Society. Ramsden's secrecy about his methods, although standard in the commercial world, violated the norms of behavior at the Society where disinterested scientific inquiry and the free dissemination of knowledge were still valued above all else.

Figure 7: Jesse Ramsden, "Circular Dividing Engine" (side view); image courtesy of the Smithsonian Institution, Washington, DC.

4.3 Pyrometers

Quadrants and sextants made of a uniform material, whether divided by hand or machine, have one rather remarkable property. Their angular division is not affected by expansion due to an increase in temperature, since as the arc expands, so, too, does the radius, keeping the angles invariant. With many other instruments, however, changes in temperature can radically alter their geometrical shape and hence produce variations in their performance. For

Figure 8: Jesse Ramsden, "Circular Dividing Engine" (top view); image courtesy of the Smithsonian Institution, Washington, DC.

example, as we have seen in Chapter 3, clocks with simple pendulums run faster in the winter and slower in the summer because their pendulums are shorter in the colder months and longer in the hotter months.

George Graham hypothesized that he could remedy the problem of irregular clocks by finding "two Sorts of Metals, differing considerably in their Degrees of Expansion and Contraction." After "several Trials," he decided that there were no two solid metals whose degrees of expansion were sufficiently different over a range of normal room temperatures. He kept this problem in mind, however, when, six years later, he "made Trial of Quicksilver" and found that "the extraordinary Degree of Expansion... immediately suggested... the Use that might be made of it." Graham then set about constructing a brass pendulum with a pillar of mercury (quicksilver) in a glass column attached to the pendulum. He adjusted the amount of mercury in the column until the pendulum beat uniformly over a range of temperatures. This clock was placed next to an identical one with a simple brass pendulum, and for the next three years and four months their times were compared against each other and against the absolute time given by the transits of the fixed stars. Graham concluded that the mercury pendulum's beat varied over the year less than "a sixth Part of the other Clock" and that it was more exact than an ordinary clock because

> ...as Heat lengthens the Rod of the Pendulum, at the same Time it increases the Length of the Pillar of Quicksilver, and its Centre of Gravity is moved upwards... [and] the Distance between the points of Suspension, and the Centre of Oscillation of the Pendulum, will be always nearly the same.[257]

While this design was not robust enough to travel to the Arctic with Maupertuis, or to the Tropics with Campbell, Graham was very pleased with it, and it is prominently displayed on the table behind his left arm, in the portrait by Thomas Hudson that was commissioned by Lord Macclesfield (Figure 9). Once Graham had demonstrated the principle that two different materials could, when correctly arranged, counteract each other's expansions and contractions to produce an invariant pendulum, other instrument makers built machines to measure the expansive properties of specific pieces and classes of matter.

Figure 9: George Graham, painted by Thomas Hudson, engraved by J. Faber; image by permission of the Royal Society of London, London.

In 1736 John Ellicott, a watchmaker and F.R.S., showed "a curious Engine...for measuring the different Expansions of different

Metals under the same degree of heat" to the Royal Society but published no measurements.[258] The whole instrument, shown in Figure 10, is made of brass, fixed to a heavy mahogany base. The bar being tested expands as it is heated and, via the spring and pulley mechanism, moves the pointer on the bar. A movement of one degree on the dial corresponds to an expansion in the bar of $1/7200^{\text{th}}$ of an inch.[259]

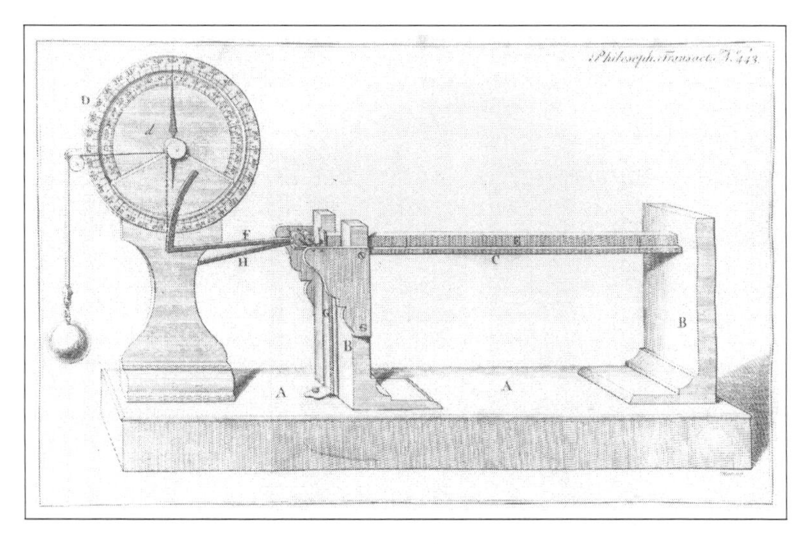

Figure 10: John Ellicott, "Pyrometer," in Ellicott 1736; image by permission of the Princeton University Library, Princeton, NJ.

Ellicott would write more about this instrument and the practical fruits of his researches.[260] Having been asked to copy a pyrometer of Petrus van Musschenbroek, he decided to build an improved version. In the process, he discovered that brass expanded three measures of length for every two measures of length of iron.[261] With this knowledge, he built the pendulum shown in Figure 11. The iron rod has a brass rod screwed onto it. The brass ball is free to move up and down the pendulum rod. When the pendulum is heated, the brass rod expands further than the iron, thus pushing down on both levers. Since the iron bar will expand downwards two measures, the ball needs to be lifted upwards two measures in compensation. This is achieved by the brass rod expanding sufficiently to move the adjustable levers which themselves move the brass ball. Ellicott showed this pendulum to the secretary of the Royal Society, who in turn showed it to Graham. Graham objected that the brass rod fixed to the iron rod would expand or contract in jerks. Further work satisfied Ellicott that this was not the case and by 1752 he was able to show his pendulum to the Fellows and give them "the Prints of the Pendulum and the several contrivances described in this Paper."[262]

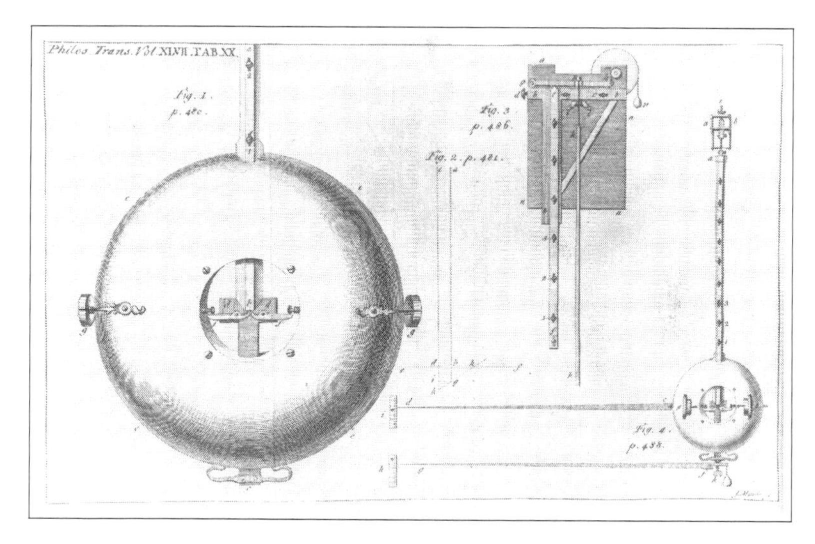

Figure 11: John Ellicott, "Pendulum," in Ellicott 1752; image by permission of the Princeton University Library, Princeton, NJ.

Ellicott, to burnish his reputation for scientific investigations, chose to have this invention appear in the foreground of the very fine portrait of himself by Nathaniel Dance (Figure 12). Indeed, as was the case with Graham, it is the only instrument displayed in the portrait. However, Graham's objections surrounding frictional forces turned out to be correct. The pendulum worked more in theory than in practice. It was to be John Harrison who used the same principles of the differential expansion of metals to construct a pendulum—his justly famous steel and brass "gridiron" design—that was not particularly affected by temperature variation.

Figure 12: John Ellicott, painted by Nathaniel Dance, engraved by Robert Dunkarton; image by permission of the Royal Society of London.

Two years later, John Smeaton demonstrated for the Society a newly designed pyrometer.[263] Made of brass, the entire instrument was immersed in a water bath while the measurements were made. As in Ellicott's pyrometer, the material being measured is placed on a bar and the material's expansion moves part of the instrument which touches gently against a micrometer connected to a measuring dial. Smeaton claimed an accuracy of one-quarter-of-an-index division, corresponding to an expansion of $1/2345^{th}$ of an inch. He measured absolute expansions by using a bar of wood as the reference length. His procedure was very careful at this stage. He wrapped a steel-tipped bar of wood in linen and measured its length at 30-second intervals. Since the bar's length expanded by a geometric progression, he could extrapolate back to the time when the bar was first placed into the hot water to calculate the length of the wood before the heat of the water bath had begun to expand it. The other Fellows were duly impressed by such mechanical exactness:

> [A] bar of Steel... was taken out, and held in the hands of one or two persons, for about a minute, by which Small Degree of Heat it was so lengthened, as to produce an Alteration of between five or six divisions on the Index. And further to shew the accuracy of the micrometer... a single piece of common writing Paper, was placed between the bar and the Instrument, which produced an Alteration of 17 [&] 1/2 divisions, and since a quarter of one of these divisions occasions a Sensible change, it appears that an Alteration in the length of the bar, not greater than the 70th part of the thickness of common writing Paper may be judged by this instrument.[264]

Thus Smeaton could measure the mechanical effect of a warm hand, and contemplate the fractional thicknesses of a piece of paper. He compared his results to Ellicott's and noted their proportional agreement. Graham's, Ellicott's, and Smeaton's papers show that instrument makers were able to gain certain knowledge about the properties of the brass and steel they worked with. They had faithfully followed "the incomparable Lord *Bacon* [who]... assured his disciples, that... by a proper induction from sober and severe experiments... [they] should in the end command Nature."[265] As a result of their researches they commanded Nature sufficiently to build clocks and watches of an astounding accuracy. The most famously exact clock maker of them all, John "Longitude" Harrison, worked somewhat separately from the above men, though he received invaluable assistance from George Graham (including a loan).[266] He was not a clubbable man; one of his more sympathetic biographers, Humphrey Quill, variously calls him "prickly ... egotistical... stern and testy."[267] In chasing the immense longitude prize he developed a paranoia towards those he suspected of trying to deny him his reward. While never wishing to be elected a Fellow, (probably because of his social awkwardness and suspicious nature), "it was due to the constant help of the Royal Society, that the Board of Longitude maintained their faith in John Harrison over the barren years."[268]

The Royal Society recognized Harrison by awarding him the Copley Medal for his detailed experiments on brass and steel, to make his heat invariant gridiron pendulum and bimetallic spring.[269] In bestowing the medal upon Harrison in 1749 (before he had completed the watch that fulfilled the conditions of the Longitude prize), Martin Folkes, the President of the Society, eulogized that after an "almost infinite number of experiments," Harrison made "a compound pendulum, whose centre of oscillation was... at the same absolute distance from the point of suspension, through all the several variations of heat and cold."[270] Sixteen years later, the Society bestowed vicarious honor on Harrison by electing his son, who had overseen the sea trials of his chronometers, a Fellow in 1765. The recommenders for the son's election said what could have been equally well applied to the father: that he was "skilled in mechanics, and several other useful parts of mathematical learning," as well as being "well known to the learned in this kingdom and several parts of Europe, for his careful experiments for the discovery of the [finding of] longitude at sea."[271]

4.4 Vitrometers

No physical theory of the eighteenth century could predict the coefficient of expansion of a particular metal; the only option was measurement. A similar problem existed in optics: as there was no known way of calculating the dispersion of light by a particular kind of glass, one could not design an achromatic lens until one had first measured the dispersive power of each piece of glass that would be used in the lens. For mathematicians like Euler, whose "rational mechanics depended on knowing matter *a priori*, without experiments," such issues caused great philosophical difficulties.[272] But for instrument makers, the fact that different materials had changeable and sometimes unpredictable qualities, cohered well with their everyday experience and suited their interests in the making of instruments of measurement. Indeed, for opticians like John and Peter Dollond, variations in the dispersive power of glass represented an opportunity to best their rivals.

There are no extant documents that decisively indicate how the Dollonds set about measuring such variations. But there is a detailed account by Father Roger Boscovich, the peripatetic Jesuit, of an instrument that may well be similar to the one they used. When Boscovich visited London in 1760 and 1761, he was elected to the Royal Society and "personally visited Mr. Dollond." As a result of this visit, "from then on, there was a new line of activity for Boscovich: the calculation of achromatic lenses and the design of new instrumentation for the measurement of optic parameters of glass."[273] Boscovich championed John Dollond's methods and acknowledged his intellectual debt to the London optician, noting that in his own optical works he had followed the "first principles of Dollondian science."[274]

The key to executing these principles was a vitrometer to measure the dispersive power of various pieces of glass. Boscovich's vitrometer is depicted in Figure 13.[275] The instrument rests on a brass compass which can be opened and closed, thereby moving the two small pieces of glass past each other. When the compass is shut the two flat sides of the pieces of glass are parallel; when the compass opens, the flat sides become angled and form a prism. This angle can be varied (and measured by the vernier) hence making the two pieces of glass a variable prism. The piece of glass whose dispersive power is being measured is placed against the variable prism and light is directed through the ensemble. This arrangement provides another way of replicating Dollond's experiment in Figure 5, in which two prisms are placed next to each other to see what prism angle will remove the chromatic aberration. In this at least, the influence of John Dollond is clear. The advantage of Boscovich's vitrometer is that the angle of one of the prisms may be varied until it removes the aberration. Each of the two pieces of glass to be fitted together as a compound lens are placed in turn on the vitrometer: their dispersive powers are measured with respect to the glass and hence with respect to each other.

Boscovich did not make the vitrometer himself: "after its form came to my mind in Venice in 1773, it was made perfect by the famous telescope maker, Dominico Selva."[276] Whether Boscovich came up with the design of the vitrometer on his own accord, or based it on something he saw in the Dollond workshop during his visit to London, is not possible to say; the Dollonds were, as we shall see in Chapter Five, very secretive about the crucial parts of

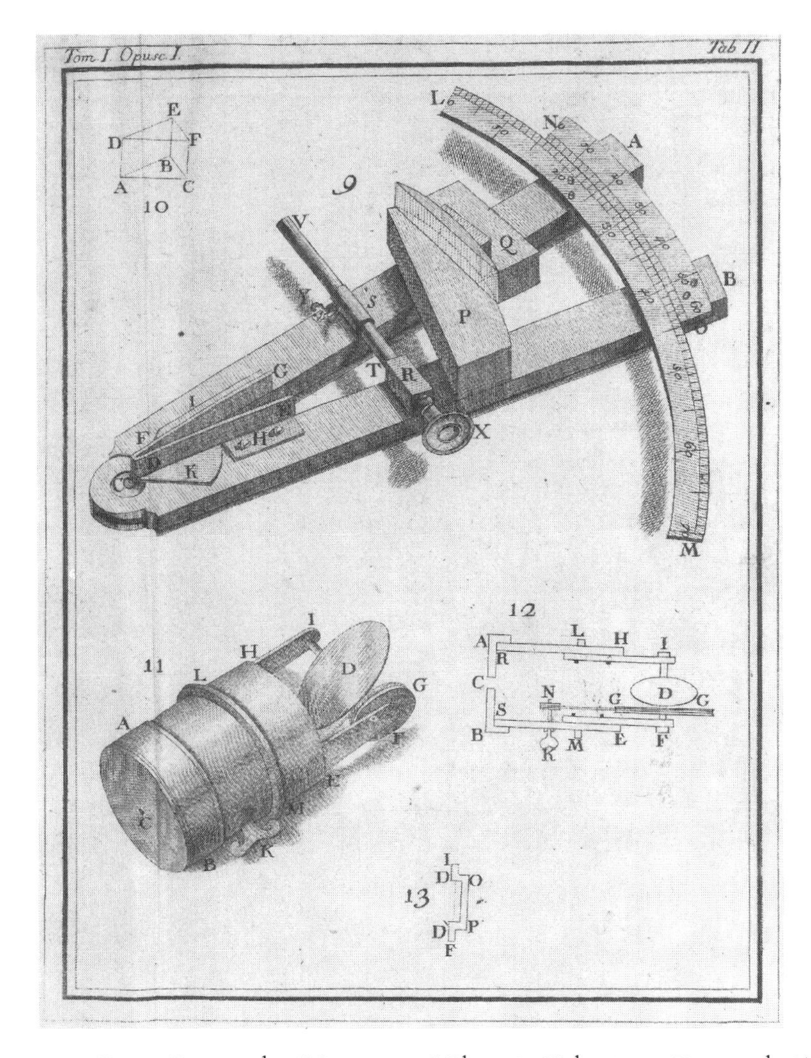

Figure 13: Roger Boscovich, "Vitrometer," Plate 2, Vol. 1, in Boscovich 1785; image by permission of the Houghton Library, Harvard University, Cambridge, MA.

their business. We do know, however, Peter Dollond took a very close interest in the Selva family, no doubt because they claimed they were making achromatic telescopes that were just as good as those the Dollonds made.

To find out more about the Selvas, Peter Dollond had translated for himself the "Fourth Dialogue" from the *Six dialogues on practical and theoretical optics*, written by Dominico's son, Lorenzo Selva.[277] Although it cannot necessarily be taken as an indication of the actual methods of the Dollonds, the "Fourth Dialogue" at the very least does tell us what kinds of optical practices and theories Peter Dollond knew about, and indicates how keen were other opticians to copy Dollondian methods, and hence bring to market telescopes they could claim were as good as those of Dollond. The work is set in the form of a Galilean trialogue and a partial version is transcribed in Appendix B. The main speaker is the author, Professor Selva, Public Optician of Venice, who may be taken to represent Lorenzo's views. The other two are the Reverend and the Count; the former being more quick-witted than the latter but both men of good will towards the Professor. The first part outlines the invention of the achromatic lens, the second the difficulties of making good flint glass, and the third the operation and usefulness of the vitrometer that Selva made for Boscovich. Like the Dollonds, the Selvas clearly understood the need to measure the dispersive power of every piece of glass that went into a compound refracting telescope lens; indeed, Lorenzo makes public this measurement practice to assure the reader that a Selva instrument is as good as a Dollond.

Venetian (crown or soda) glass was the best in Europe in the seventeenth century. In attempting to better it, some English glass manufacturers made a new kind of glass containing lead.[278] The high dispersive power of this English flint glass turned out to be crucial to the successful operation of Dollond's achromatic lenses. In a humiliating reversal, Venetian opticians were forced to import whatever piece of English flint glass they could get to try and make achromats; Selva, for instance, reports having obtained twenty pieces, of which only seven were free of defects.[279] Peter

Dollond was crucially interested in improving the quality of glass for his own business but to little avail.[280] In 1771 he sat on the Chemistry Committee of the Society of Arts to evaluate the entries for the £60 prize for a batch of flint glass weighing more than twenty pounds. Though he and Edward Nairne made a partial award to Richard Russell, they decided that none of the entries was sufficiently free of "waviness" to be worthy of a full prize.[281] Because the glass was so important and yet so often defective, the Board of Longitude even offered a prize of £1,000 in 1779 to anyone who could reliably manufacture defect-free flint glass. The prize was never awarded.[282] Thus, Selva bemoans the "great scarcity of good flint glass in England" and the failure of efforts to "purify it and render it free" from threads and streaks.[283]

Even when using the best flint glass available, it was not enough for a telescope maker merely to copy the dimensions of a Dollond telescope and its lenses. Because each piece of glass varied so much in its refractive and dispersive power, Selva explains, it was "necessary that the artificer understand the refractive and dispersive power as well of flint as of the common glass which he would unite with it; since that a vessel of flint-glass taken out of the same furnace will have one (power) and another (vessel) a... different (power)." Each pot, or even each piece, of glass had properties that could not be specified beforehand.[284] Careful measurement with a device such as Boscovich's vitrometer was therefore crucial if a telescope made by Selva in Venice were to be as good as one made by Peter Dollond. And "tho' to some people the price of a Dollondian telescope seems immense," Selva argues, its achromaticity, attained through a meticulous process of "combinations and calculations," made it well worth the price.

Chapter 5

Credit and Discredit

John Dollond's achromatic lens was hailed as the "greatest improvement in optical instruments" since the time of Newton, an invention "of the first rank," and "a vastly more manageable and useful instrument" than any previous telescope lens.[285] At the Royal Society, Dollond was acknowledged both as the inventor of the artifact itself and as the originator of the major theoretical discovery it embodied. James Short credited Dollond with having single-handedly created a "theory of correcting the errors arising from the different refrangibility of the rays of light in the object glasses of refracting telescopes." The President of the Royal Society, the Earl of Macclesfield, emphasized, in his Copley Medal address Dollond's meticulous application of the experimental philosophy so favored by the Society: "this improvement he was led to by the result of a long course of Experiments." And Nevil Maskelyne went so far as to declare that "Mr. Dollond had made a theory" that was "perfect."[286]

However, so valuable an intellectual property as Dollond's new invention attracted jealous attention from others in the general European optical community. In the 1760s, in a series of intriguing patent cases, a group of London opticians and a shadowy English country gentleman named Chester Moor Hall asserted their rights to being the original inventors of the achromatic lens. And in the same decade, Euler laid claim to the theoretical parts of the discovery.[287] Although Peter Dollond managed to assert his father's right to the patent that described the invention, he had no means of silencing the Eulers of the world, who lay beyond his reach. By 1789, when Peter Dollond wrote two impassioned letters to the Royal Society defending his father's reputation, no one at the Society was willing to step forward and vindicate the memory of John Dollond, a Copley Medal winner and former distinguished Fellow. Indeed, Dollond's own son-in-law, Jesse Ramsden, was among his most vigorous detractors. This posthumous dismantling of Dollond's reputation reflected the increasingly tenuous position of instrument makers at the Society, where the scientific contributions of shopkeepers such as Dollond—not to mention Ramsden, the last of the great eighteenth-century makers to be elected a Fellow—were increasingly devalued.

5.1 Opticians

From 1758 until his death in 1761, John Dollond held—at least in London—unchallenged credit for the invention of the achromatic lens that he and other instrument makers made and sold. Had matters remained thus, Dollond's reputation as the undisputed inventor of the lens would no doubt have remained intact. Other London opticians would have continued to manufacture achromatic lenses as they saw fit, happy to acknowledge Dollond's success in publicizing the new invention throughout Europe and hence creating a demand that they could profitably satisfy. After his death, however, Dollond's son Peter decided to lay claim to commercial property in the invention: a patent of 1758 that was never enforced during his father's lifetime. To defend themselves from the monopolistic Peter Dollond, who wanted them to stop making achromatic lenses altogether, members of the London optical community produced, from obscurity, Mr. Chester Moor Hall, whose indirect claim to the invention has clouded John Dollond's reputation ever since.

John Dollond's patent of 1758 states (after a preamble about the great utility of the invention to the nation and to the navy) that:

> ...he hath by Application and Study and at a Considerable Expense at last Invented and brought to perfection a new method of making the Object Glasses of Refracting Telescopes by Compounding Mediums of Different Refractive Qualities whereby the Errors arising from the Different Refrangibility of Light as well as those which are produced by the Spherical Surfaces of the Glasses are perfectly Corrected. ...And...he is the first and sole Inventor thereof.[288]

Dollond claims that his lenses are free from two problems–chromatic and spherical aberration—and that he is the first optician to have achieved both things. But what did the lens look like? The patent contains no diagrams, which was not unusual for the time. A diagram did appear a year later in Benjamin Martin's optical textbook, however; it depicts the same arrangement that Rolf Willach has found in a very early Dollond lens and is consistent with the vague description of the lens in the 1758 paper and patent.[289] It is a doublet in which the flint concave lens faces towards the object and all four radii are spherical, as in Figure 14 (hereafter Flint Forward Spherical).[290]

Dollond could not afford to patent the design on his own. Obtaining an eighteenth-century English patent cost approximately £70 in charges and stamp duties to a variety of government offices, and often as much again in other fees to officials and solicitors who helped navigate the application through those offices in the

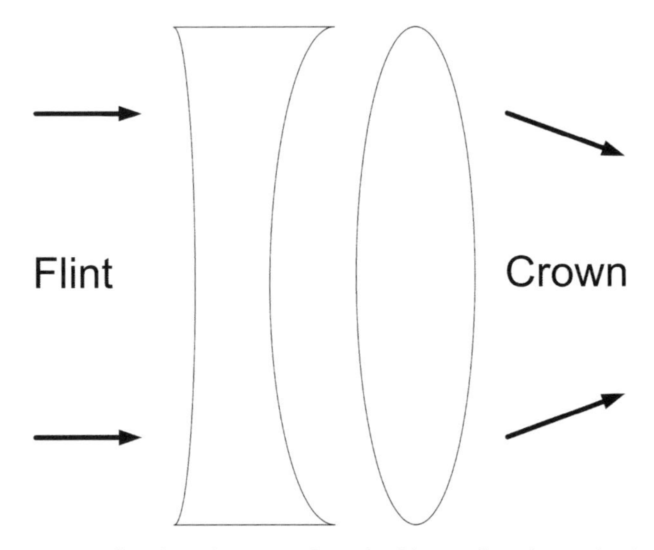

Figure 14: Dollond's achromatic lens doublets—flint forward spherical

correct order. Following upon that were the extensive and open-ended expenses of defending the patent if it were (as it usually was) infringed.[291] This kind of expenditure was completely beyond someone who worked as a skilled tradesman; additional capital had to come from profits, loans, stockholders, or partners. It was to this last source of capital that Dollond turned, assigning a half-interest in the patent to another optician, Francis Watkins, who paid the substantial costs associated with obtaining it.[292] Since Watkins put up the money, it is possible that the patent was more his idea than either of the Dollonds. Watkins may well have realized that, with their new lens, John's international reputation, and Peter's commercial brilliance, the Dollonds posed a very grave threat to his own refracting lens business and it was better to co-opt them into a partnership than face them down in the marketplace.

But why, having got the patent, did neither Dollond nor Watkins try to stop other London opticians who quickly began to make the Flint Forward Spherical achromats? Perhaps, with the difficulties of making the new lens, Dollond & Son, and Watkins could not meet the demand on their own and did not object to other opticians filling it. In any event, at John Dollond's death in 1761, his half-share in the patent descended to his children, including the leader of the business, his son Peter. John Dollond died intestate and it is not clear how the property interest in the patent was precisely distributed among his children. Jesse Ramsden received a portion of this half-share in the patent when he married Peter's sister, Sarah.[293] The other half-share of the patent remained in Watkins's control until 1763, when he agreed to sell it to Peter for £200 at about the same time as their partnership was terminated due to a bitter dispute over how to divide up the profits associated with the manufacture of the lenses.[294]

It was this dispute, in the first instance, that motivated Peter Dollond in 1763 to sue Watkins, who kept on making and selling achromatic lenses, thus infringing the patent. At the same time, Peter Dollond also decided to sue every other optician who was making achromatic lenses, some of whom had been doing so since soon after 1758. Besides his animosity towards Watkins, he had another very good reason indeed to try and monopolize the trade. By 1763 Peter Dollond had a powerful new design for an achromatic lens that had been developed a few years earlier, either by him or his father or the two in concert. This new design, shown in Figure 15, is probably from 1759 or 1760, since it postdates the 1758 patent and first appears in the 1760 Passemant copy in the Louvre. Willach has called this achromat "brilliant and simple" and exclusive to the Dollonds. The crown convex lens is now facing the object, and the outer radius of the flint concave lens is aspherical (hereafter Crown Forward Aspherical). Until Willach's meticulous investigation, one important feature of this lens has remained generally unknown.[295] With a very subtle amount of polishing in the outer margins of the flint lens, the spherical aberration of the entire lens was removed. The process took a few minutes and did not change any of the other properties of the lens; particularly its focus and achromaticity. This, with a few modifications, remained the classic "English" design for achromatic doublets into the 1820s.[296]

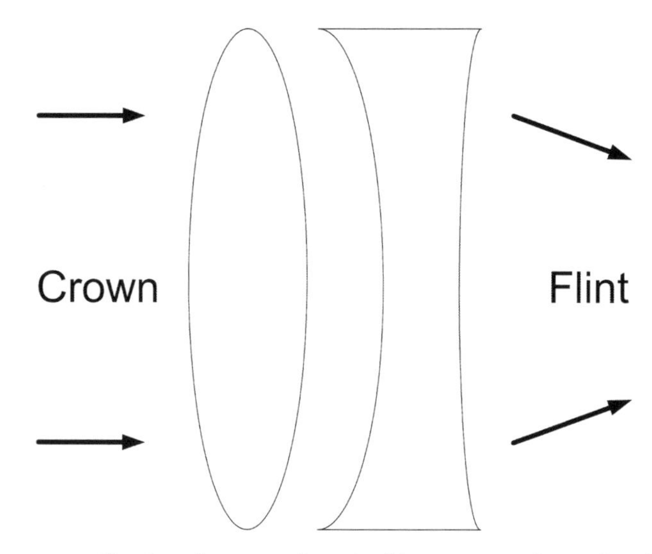

Figure 15: Dollond's achromatic lens doublets—crown forward aspherical

For Peter Dollond, the Crown Forward Aspherical design—which was never described in print, and whose novelty lay in its solution to spherical aberration—had several tremendous commercial advantages. He could now make more achromats at less cost; thus, if he could control the supply (and hence maintain the customary price), his profit per lens would increase. And if he were to monopolize the trade, perhaps he could even raise the price, as indeed he did following his successful patent suit.[297] Even before the patent trial, Peter Dollond sold achromatic refractors "at a much higher rate than that what they have always been sold at by others in the Trade," according to one of the opticians sued by him, Christopher Steadman, so the delicious prospect of selling more lenses at even a higher profit margin proved irresistible.[298] So long as the new design remained a trade secret all was well. He did not need to enforce his patent. But with the possibility of the disgruntled Watkins spreading the secret around London after 1763, Peter Dollond knew he could be out-supplied and undercut by the rest of the London optical community. So he took the risky and expensive step of calling in the lawyers. To the modern reader, this tactic reeks of dubious morality and practicality, since the 1758 patent grant could only cover the earlier Flint Forward Spherical achromatic lens, not the later Crown Forward Aspherical achromatic lens. But as we shall see, this distinction was not significant in the London law courts of the 1760s. Rather than questioning what particular kind of achromatic lens it was that the patent covered, the legal proceedings focused on whether or not Dollond's lens was the first commercially available achromatic lens of any kind.

Peter Dollond and his lawyers had literally no precedents upon which to estimate their likelihood for success, because "the first major patent case at common law since the early seventeenth century was Dollond's."[299] In the intervening century and a half, a patent had become a rather different thing than before, so the much earlier cases were of little use. At the beginning of the seventeenth century, letters patent were issued by the Monarch either to give the holder, usually a court favorite, a monopoly over some particular item of trade or to encourage the importation of technology from abroad. The patent's evolution from an early seventeenth-century court privilege to a late eighteenth-century commercial reward for innovation is deftly outlined by Christine Macleod, who argues that the "purchase of a patent was a commercial transaction," whose "aim might have been to protect and exploit an invention; or it might have been to impress potential customers or investors; to escape the control of a guild, or to replace a guild's protective cloak, when that began to grow threadbare."[300] As we shall see, Peter Dollond hewed strongly to all of these aims but the last.

John Dollond's original patent application promised the authorities that Dollond would "particularly describe... his said Invention and in what manner the same is to be performed."[301] Although a monopoly was granted to the patent holder for fourteen years, on the presumption that the invention was clearly described and original, in reality the patent was simply registered and not checked for either characteristic. Therefore those who chose to could argue that the patent they were infringing had neither clarity nor originality, and the task of deciding whether a patent was valid and exactly what invention it described fell to the courts rather than to the enrolling body. The financial burden of protecting a patent was borne by the patent's owner until (if ever) he was successful in his suit. The complexity and expense of litigation and the uncertainty of the outcome meant that many patents were not defended by their owners despite flagrant infringements.

Peter Dollond attacked Watkins and his new partner, Addison Smith, with a suit in King's Bench in 1763 and another in 1764. These were the first of his twelve lawsuits against seven different opticians, every one of which he won. Although the damages Watkins and Smith were ordered to pay were very light (a few shillings), they were also instructed by the court to stop making any more achromatic lenses of any kind.[302] After these initial setbacks Watkins tried another tactic to defeat Peter Dollond. He got himself elected Master of the Spectaclemakers' Company and, with his guild's support, tried to derail the judicial proceedings by appealing to the Privy Council.[303] The Privy Council, an ancient body of advisors to the Monarch, traditionally heard petitions to the King asking him or his servants to intercede on his subjects' behalf. In theory, the Council could curtail the actions of a judge or invalidate a patent.

The general complaint of the petitioners to the Privy Council was that the achromatic lens was so successful that virtually no one was now buying reflecting telescopes. Since telescopes were the most valuable branch of their business, they would be ruined if they could not make achromatic refracting telescopes, and "the said Peter Dollond under Colours of the said Grant [would thus]. . . Ingross to himself the Exclusive Right to almost the whole Branch of making Telescopes in General." Nearly everyone in the London optical community (Peter Dollond and his brother-in-law Jesse Ramsden being obvious exceptions) was worried about and outraged by Peter Dollond's behavior. Only a few of those who signed the petition confronted immediate ruin if Dollond's legal actions against them were successful, but the rest would face a problematic future if Peter Dollond could monopolize the refracting telescope trade until 1772. By 1764, the entire Spectaclemakers' Company had become used to making achromatic lenses, and they saw no reason why they should not continue doing so. Their argument was general, against any patent that might cover so large a part of a given trade, rather than specific about the validity of Dollond's patent. Just as Peter Dollond had no precedent to indicate his likelihood of success in prosecuting a patent infringement case, so too the Spectaclemakers' Company had no previous experience in having a large part of their livelihood legally and suddenly removed from them.

Such a general complaint, however, had very little chance of success. Firstly, the Privy Council no longer regulated patent cases, so it could not answer objections about the patent itself. Secondly, the English state was generally unsympathetic to the petitions and claims of guild members: "unlike Burghley or Coke, eighteenth century Chancellors and judges showed little respect for the conventions of craft fellowship, guild solidarity or protection of traditional ways of working if they obstructed the path of innovation and efficiency."[304] By the mid-eighteenth-century, guilds were in severe decline and had so lost their original function of regulating working practices that many instrument makers were members of the Grocers' Company. Attacks on Guild Companies by the Courts and Parliament culminated in the 1813 Act of Parliament removing the apprenticeship regulations dating from the early years of Elizabeth's reign.[305]

When the Privy Council declined to hear the petition, the opticians infringing the patent were left to face Peter Dollond again in the Courts of King's Bench, Common Pleas, and Chancery.[306] Robbed of a general defence, the specific opticians named in the lawsuits were forced to focus on the particular nature of the patent itself. They tried two tactics to argue that the patent must be invalid: it was unclear and it was unoriginal. On the first point, defendant Francis Watkins told the Chancery Court that in his opinion patents were only granted on the understanding that the invention was sufficiently clearly described, so that "after the expiration of the Term Granted such Invention may be used and Practised in General for the public Benefit."[307] While Watkins and other opticians may have thought this theory about what a patent was for would appeal to a judge and jury, they knew it to be essentially irrelevant to their own practice. They learned how to make lenses and mirrors of a new construction not from written patents but from being shown how to do so by a master practitioner.

Until the second half of the eighteenth century, patents were very unusual in the instrument-making trade in general and were actively opposed by the Spectaclemakers' Company in particular. They became more common as the influence of the guilds waned. Watkins's claim that the patent was very obscure is all the more peculiar when we recall that he had owned a half-share in it from 1758 to 1763 and may even have been the driving force behind obtaining it. While he did not write the patent (John Dollond supplied a draft of the description of the lens to the legal clerk), Watkins must have approved of its content before paying the substantial expenses associated with securing it. Indeed, as a beneficiary, he may well have appreciated the imprecise wording at the time. Nonetheless, by 1765, he professed himself shocked to find such language in a patent. Other less-partial witnesses and defendants also swore it was impossible to learn anything from Dollond's patent; among them was the mathematical instrument-maker John Bird, who thought that "the specification of the patent... was *not* sufficiently explicit to enable any workmen to make such object glasses."[308]

Dollond's patent description was admittedly cryptic; he provided neither diagrams nor the mathematical figure of the lenses nor any indication of the materials used. Such imprecision was, however, not at all unusual for patents of the time and is not mentioned as a deciding factor in surviving accounts of the verdicts. A few defendants seem to have realized that the issue of specification might have been turned to their advantage. Several argued that "such new invention must consist... in the particular formation of the Convex and Concave glasses and the mediums and substances... in which particulars the said pretend Specification is totally Silent."[309] None, however, made any distinction between the Flint Forward Spherical achromat described, however vaguely, in the patent and the post-patent Crown Forward Aspherical achromat that Peter Dollond was manufacturing and wished very much to monopolize.

Since the patent's lack of clarity seems not to have unduly worried the courts, the defendants tried another tactic: John Dollond's invention, they argued, was not original. To prove this, they dragged from his rural home a retiring gentleman who was prepared to swear that he had already invented the self-same lens thirty-or-so years earlier but had neglected to inform the world of his marvelous discovery. Thus did the public first hear of Chester Moor Hall, a lawyer and country gentleman from Essex, about whom remarkably little else is known. Two of the most careful scholars who have studied this case call him "semi-legendary."[310] No one has uncovered any documents prior to these patent trials, published or unpublished, in which Hall or anybody else claims to have invented the achromatic lens. The most recent supporter of his claim, Rolf Willach, admits that "we have very little knowledge of him... Hall never published any information about his research on the achromatic object glass. The only written documents to survive him are some signatures on judicial files."[311]

Since Hall was not selling any lenses and hence not infringing the Dollond patent, he was not subject to legal proceedings; thus there are no affidavits from him. His story is told in the court records entirely through the affidavits of the defendants, who claimed that Hall, after having London opticians make lenses for him privately in the 1730s and 1740s, had given enough information about his new lens to the London optician James Ayscough in 1752 to have a crown and flint doublet made. Ayscough then turned to another optician, William Eastland, to execute the order. A year later Eastland made and then displayed in his shop a few telescopes containing this lens.[312] Thus on two grounds—by causing a supposedly achromatic doublet to be privately made in the 1730s and 1740s, and by making possible the public display of one in the early 1750s—Hall was the prior inventor of the achromatic lens and the Dollond patent should be invalid.

But awkwardly for the opticians, in every instance they had declined to make the lenses on their own initiative. The artifacts were only ever made at Hall's urging. The "invention"—in which the defendants demonstrated an interest only after the appearance of Dollond's paper in the *Philosophical Transactions* and the registration of the patent—had languished in obscurity, uncopied and unsold, prior to 1758. Francis Watkins, in particular, was caught yet again in a tricky situation. Why had he sought the patent in the first place, if he believed it to be fraudulent? And, even more damaging, why had he accepted £200 from Peter Dollond for his half-share of the patent in 1763? Watkins explained away such apparent contradictions by saying that he had accepted that sum because he estimated the patent had only a small chance of being upheld in court. If it were judged valid it would be worth very much more. Watkins could not deny that he had learned to make achromatic lenses from his erstwhile partner, John Dollond, rather than from a shy Essex lawyer. Addison Smith, similarly, by virtue of his partnership with Watkins, had learned (indirectly) from Dollond and not from Hall.[313] This bears repeating. Skilled members of the Spectaclemakers' Company who had great difficulty making the new lenses under Hall's direction from the 1730s onwards had little trouble copying Dollond's lens in the late 1750s and on into the 1760s. Furthermore, the few Hall lenses—whether or not they were actually achromatic—remained so obscure that Dollond knew nothing about them when he began his correspondence with Euler in the 1750s. The failure of the opticians to commercialize Hall's supposed invention led one presiding judge, Lord Mansfield, to instruct the jury that "what Mr. Hall had done must be considered as nothing; for if a man had ever so useful an invention, and he kept it locked up in his Scrutoire, it was the same thing to the world, as if he had never known it."[314] The jury agreed, and found John Dollond's patent to be valid. They also awarded substantial damages to Peter Dollond in this and other cases and ordered the infringing opticians to stop making the lenses until the patent expired in 1772.[315] Thus Peter Dollond became the sole manufacturer of achromats (all of the Crown Forward Aspherical design) during that period.

By the time the patent expired, Peter Dollond had acquired so much reputation, capital, manufacturing experience, and control over the crucial flint glass supplies that his dominance of the refracting telescope trade—or at least the astronomical portion of it—continued on into the early nineteenth century. As a consequence, public and private telescope collections containing late eighteenth-century astronomical refractors invariably have Dollond's and very few of anyone else's.[316] Yet the patent trials, though commercially lucrative for Peter Dollond, contained a ticking time bomb that went off to such an extent in 1789 that afterwards John Dollond was never again credited for the original invention of the achromatic lens.

5.2 Mathematicians

In the same decade that the London optical community attacked Dollond's reputation as the inventor of the achromatic lens, European mathematicians laid claim to the theory behind it. In 1762, Euler wrote a letter, later to be published in his *Letters to a German princess*, stating that he had "given the first ideas" for these lenses "several years ago, and since that time the most cunning artists in England and France have laboured to execute them." Dollond, Euler said, had succeeded in making good telescopes only after entering

> ... into a correspondence with me about this matter; but since it principally concerned the surmounting of various great obstacles of practise, with which I have never been involved, it is as well to abandon to [him] the glory of discovery; it is the theoretical part alone that is mine and which has cost me profound researches and the most painful calculations; the very sight of which would terrify your highness.[317]

It seems safe to assume that not only German princesses but also French instrument makers were terrified by Euler's calculations, for they made no use of them.

In fact, as we saw in Chapter Three, Euler's correspondence with Dollond was not about practical matters alone. In the 1753 letter that he wrote to Euler on behalf of the Royal Society, Dollond had checked Euler's mathematics and found the calculations correct, but he had argued that one of Euler's initial premises did not accord with Newton's theory. In replying to this letter, in the same year, Euler called Dollond "un habile Mathématicien Anglois."[318] By 1762, however, he was no longer referring to Dollond as a mathematician, but rather as a "très habile artiste en Angleterre"—a subtle but important demotion.[319]

Nor was Euler accurate in stating that Dollond, in making the first achromatic lens, had merely executed Euler's own theory. Firstly, since Euler's notion that the human eye is achromatic is in fact incorrect; Dollond could not have produced an achromatic lens had he in fact copied Euler's principles. As one editor of Euler noted: "object-glasses of this kind, even if executed in the most correct manner, are incapable of producing the effects which our author expected from them."[320] And secondly, Euler himself initially demonstrated a poor understanding of how Dollond's lens actually worked. At first, he refused to believe that different glasses with the same refractive power could have widely differing dispersive powers. Dollond's lenses, he initially averred, were not really achromatic but merely appeared so, whether because crown glass filtered out the red rays, because Dollond had stumbled on a happy curvature for his glasses, or because the eyeglasses of his telescopes partially removed color.[321]

Not until 1765, seven years after the publication of Dollond's paper, did Euler finally admit that Dollond was correct.[322] Euler was resistant to accepting Dollond's discovery because it was disturbing to an Enlightenment mathematician.[323] Indeed, it "appeared" to Euler "to shock good sense" and to be "an astonishing discovery" that the dispersive power of any given piece of glass could not be calculated *a priori* but could only be measured.[324] Measurement, not pure mathematics, was the sole and only guide to this particular problem. Euler was not alone in attacking Dollond for being a mechanic rather than a mathematician. Clairaut, while acknowledging Dollond's claim to the lens, thought that Dollond had not provided a mathematically respectable theory for his invention. He acknowledged that Dollond's lenses were achromatic, but roundly criticized him for not describing the theory of their construction in a manner acceptable to the French academic community, declaiming that one

> ... cannot even find in his paper the kinds of results that geometers usually see fit to communicate to assure us of the correctness of their discoveries... Without the help of such a theory, it is certain that we cannot make telescopes equal in value to those of Mr. Dollond, except by a servile imitation of them. Moreover one cannot be sure of success when one chooses to take this humiliating path, since it would be necessary to be assured that the materials one uses have exactly the same refrangibilities as those of Mr. Dollond.[325]

We see here Clairaut's deep frustration. Dollond violated the norms of mathematical communication because he valued his commercial secrets more. Even if he had revealed how he arrived at the geometrical shape of his lenses, unless the copier also realized the necessity of measuring the refractive and dispersive powers of the glass, the copy would be a failure. Joseph Priestley believed that Clairaut and other mathematicians never entirely grasped the importance of these crucial measurements:

> ... their geometrical theorems were too general, and their calculations too complicated for the use of workmen. ... [Of] the most consequence was the want of an exact

> method of measuring the refractive and dispersing powers of the different kinds of glass; and for the want of this, the greatest precision in calculation was altogether useless.[326]

What Clairaut saw as a "humiliating" and "servile" task—measuring materials to establish their physical properties—was for Dollond the key to his success.

5.3 Intellectual Heirs

Euler's notion that John Dollond had merely executed Euler's own theory resurfaced again in a 1783 eulogy of Euler.[327] This, and other writings along the same lines by Jérôme Lalande and Jean-Dominique Cassini (Cassini IV), so incensed Peter Dollond that he wrote a long and intemperate letter to the Royal Society in 1789 defending his father's reputation.[328] Since Peter was not F.R.S., the Astronomer Royal, Nevil Maskelyne, communicated it to the Society on May 21, 1789. The letter argued that Euler proposed a lens designed like the eye. Had it been executed it would not have been achromatic, and it certainly bore no similarity to the lens John Dollond made. To John Dollond alone belonged the honor of overturning Newton's Experiment VIII and establishing the principle that dispersion was independent of the mean refraction.

The Royal Society, bridling at his combative and controversial tone and perhaps disapproving of his aggressive defence of his father's patent, did not publish Peter Dollond's letter, forcing him to do so on his own accord.[329] Nor did any Fellow offer to defend John Dollond's memory, either as a discoverer of optical principles or as an inventor of optical instruments. Instead, one stepped forward to attack it: Peter Dollond's own brother-in-law, Jesse Ramsden, who averred in a 1789 response to Peter's letter that John Dollond was not the inventor of the achromatic object glass, "it having been first discovered by Mr. Chester Moor Hall Esq. a gentleman in the profession of the Law, who lived in Essex."[330] Hall was, Ramsden recalled in his letter, a man who loved "retirement" and had little "thirst. . . for public fame." Ramsden "knew him personally several Years before his Death which happened about the year 1767," and Hall told him that "it was from considering the structure of the eye [around 1729] that he first form'd the Idea of what is now called an Achromatic Telescope."[331] Hall clearly had better luck than Euler if his false analogy to the eye really led him to a correct design for an achromatic lens.

In the early 1730s Hall, being very "jealous" about his invention and demonstrating some of the characteristics of secretive tradesmen, contracted out the grinding of the crown convex lens to one optician and the flint concave to another, giving each man (Edward Scarlett and James Mann) a set of calculated specifications for that individual piece.[332] The two happened to subcontract out the work to the same optical jobber, a Mr. Bass, who seemed to realize that the two pieces (a flint concave and a crown convex) were supposed to go together, if not that the combination was achromatic.

With the secret partially out, but not it seems attracting much attention, Hall decided to entrust the making of the entire lens to one optician. He turned to the very eminent and very busy mathematical instrument-maker John Bird. Bird ignored the commission for several years, so Hall enlisted Mann's former apprentice and short-term partner, James Ayscough, but some business difficulties intervened. Ramsden is very explicit that not a single achromatic telescope of Hall's design was ever "made public by an actual sale," though he does refer to the existence of a few achromatic telescopes privately made for Hall by Ayscough (probably in the early 1750s) and, more surprisingly, by the previously reluctant

John Bird. Ramsden claims to have received all this information either from Hall or the above-mentioned opticians.

Disputes over who invented something that later became important or lucrative are legion. The credit for making the first "telescope" itself had many claimants. There was probably a device as early as 1590, a lens or eyepiece system, well before the Dutch patent application of 1608. However, its magnification was so low that no one took much notice of it or, like Hall's lens, wanted to copy or sell it. Then the telescope became something worth claiming credit for and Johannes Zachariassen claimed his father had made the device of 1590 and was thus the inventor of the telescope.[333] Ramsden did the same for Hall for the achromatic telescope lens.

Nonetheless, Ramsden's account of Hall's actions is puzzling on several grounds. Firstly, if Hall had no financial interest in developing the lens, why was he so secretive in his dealings with the London opticians? Secondly, there is something very peculiar about the story of Mr. Bass, whose accidental possession of the two parts of the lens seems an extraordinary coincidence. Thirdly, why did no one seize upon Hall's discovery, after word of it leaked out, and begin making and selling the lenses? It is not sufficient to say that Hall's design was too difficult to execute, because Dollond's first design—the Flint Forward Spherical achromat which he allegedly stole from Hall—was fairly easily copied by the general London optical community after 1758. Finally and most importantly, were any of the various lenses that Hall had made for himself effective enough in reducing chromatic aberration to be called achromatic by eighteenth-century standards?

The earliest lens, the accidental "Bass" lens from the 1730s, almost certainly was not. Even Willach, Hall's most recent partisan, concedes that "this first attempt must have been a complete failure. It was not possible, for Hall [or] any other optician in the 18th century, to manufacture an achromatic object glass on calculation alone."[334] Thus it seems likely that Hall, who produced in court "several object glasses ground by Mr. Bass" to support his claim to prior invention, misled the court about their achromaticity.[335] And while Ramsden explicitly calls the Bass lens "achromatic," he never professes to have examined it for its optical qualities himself; instead, noting that the lens was by then in "the possession of Mr. [Addison] Smith," he merely repeats the testimony of others like Watkins and Smith, who were the unsuccessful defendants in the patent trials twenty-five years earlier. Ramsden's most powerful piece of evidence for Hall's claim to the prior invention is his recollection that his father-in-law "always admitted in his conversations with me that Mr. Hall had made the Achromatic Telescope before him." But there are no surviving documents where John Dollond himself says this; all that is left is Ramsden's second-hand account of conversations that took place four decades earlier. Ramsden is very careful to note, moreover, that John Dollond did not purloin the idea of the achromatic lens doublet from Hall: "it is but justice to say that Mr. Dollond does not appear to be indebted to Mr. Hall for the discovery." The only "hint" that Dollond received from Hall's work, he says, was a very indirect one from Mr. Bass about the strong dispersive power of flint glass; otherwise, Ramsden makes clear, Dollond's invention of the achromatic lens was made independently of Hall's, though Hall's was made earlier.

Why did Ramsden write this letter against the interests of his brother-in-law, twenty-five years after the legal conflicts and sixty years after the date of Hall's alleged first discovery? While Ramsden had marriage ties to the Dollond family, he had no direct business links by 1789; indeed, he competed with them to some extent for optical business. He had worked in the Dollond workshop in the early 1760s but was not taken into partnership with Peter and opened his own shop on the Strand in 1763. There is no evidence as to whether or not he resented this exclusion from the Dollond business. He probably received a share of the achromatic lens patent as part of his marriage settlement in 1766, so he had every incentive to keep quiet about Chester Moor Hall then.

By 1789, however, the patent had long since expired and he could afford to act like a disinterested gentleman. All the same, this dispute between London's two most eminent opticians clearly had overtones of sibling rivalry to it, and Peter Dollond managed to insinuate in his response that Ramsden was ungrateful to attack him or his father:

> Mr. Ramsden... at that time was almost continually at our house getting all the information he could both in the theory and practice of optics; for he had articled himself to a Mr. Mark Burton... a very honest man and a maker of mathematical instruments, but one under whom he could not improve himself in theory of any kind.[336]

Peter implies that Ramsden received two very valuable commodities from the Dollonds: a training in optics, which he could not get from a mathematical instrument maker, and a training in theoretical matters which he could not get from any ordinary instrument maker.

While Ramsden may have had a number of reasons not to be bound any longer to his wife's family, the immediate cause of his missive to the Royal Society was Peter's long letter defending the reputation of his father against the attacks of continental opticians. Since that letter concerned the correspondence among Euler and Klingenstierna and Dollond, in which Chester Moor Hall took no part, Peter Dollond had seen no reason to include Hall in the account.[337] Ramsden, however, thought that "in justice" Peter Dollond should have mentioned Hall, whom Ramsden believed to have been the first discoverer of the error of Newton's Experiment VIII and the first constructor of an achromatic telescope, while Dollond was merely the first to make "Achromatic Telescopes... public by an actual sale."[338] In his response to Ramsden, which was also ignored by the Society, Peter Dollond gave his own account of how John Dollond came to produce the artifact and what role, if any, Hall had to play in that story. Peter Dollond's recollection of what Hall said at one of the trials differs in one key respect from that of his opponents:

> Mr. Hall's evidence was that 'he had had the same idea about 30 years before, but on account of the difficulties he had found in getting the glasses worked according to his directions... he had laid aside those pursuits... [and] he had never made any such glasses for sale, nor had he even completed a telescope.

Nor did Hall produce "any glasses upon this occasion."[339] The Royal Society—possibly viewing the dispute as an unseemly family quarrel and clearly uneasy at choosing sides between the claims of two of its most distinguished artisan Fellows (Jesse Ramsden and John Dollond) while revolting artisans rampaged through the streets of Paris—washed its hands of the dispute.

But Ramsden was interested in more than merely justice for the long-dead Hall. During the second half of the 1780s he had been under immense pressure from the President of the Royal Society, Sir Joseph Banks, and the director of the Ordnance Survey, Major-General William Roy, to complete his theodolite in a reasonable length of time so the distance between two prime meridians (the Royal Observatories at Greenwich and Paris) could be accurately measured.[340] During the survey, Roy had no choice but to flatter or plead with Ramsden to get the remarkable instrument completed but once the work was finished he could vent his resentment, which had clearly been simmering for several years, and attack Ramsden's dilatoriness in early 1790.[341] Ramsden tried to get Banks to intervene in his favour, but failed; indeed, Banks became so angry at Ramsden that he refused to speak or correspond with him over the matter.[342] Ramsden was then forced to appeal to the Society as a whole, writing a letter defending himself against Roy's invective and reminding the Fellows

> ... but had an Instrument of this sort ever been executed before, or could it have been made else where there might have been grounds for complaint; but surely a mechanic who employs his mind, neglects a constant business, and spends his money for the promotion of Science, and after all receives abuse, has much reason to be dissatisfied."[343]

By 1795, things had cooled down enough for Banks to award the Copley Medal to Ramsden, in part for his extraordinary theodolite, emolliently recalling of the earlier events that "both General Roy's character as a man of patience and Mr. Ramsden's as a man of punctuality, underwent some severe trials."[344]

When he wrote his letter about Hall in 1789, however, Ramsden still felt beleaguered at the Royal Society. Possibly he hoped that by telling the Hall story against the interests of his commercially aggressive brother-in-law, Peter Dollond, he could demonstrate to other members of the Society that he was a lover of science more than of commerce; of truth more than of family. And there is one clue in his letter that this is so. Ramsden argues that Hall deserved credit for the invention because he had first uncovered a philosophical principle—that of "the fallacy of Sir I. Newton's experiments"—and consequently a new law of dispersion.

John Dollond, he says, only deserved credit for the invention on the basis of the court proceedings, which had revealed the legal principle "that it is not the inventor but the person making a useful invention public, who is intitled to the priviledge of a patent." Ramsden wishes to split the credit between the two men, very much to the advantage of the former: Hall, the gentleman and man of science, used his head and came up with the idea of the lens first without any thought of gain; while Dollond, the artisan and shopkeeper, secured the patent by using his hands and making significant quantities of the new lens. Only Hall's supposed accomplishment constitutes invention for Ramsden; that the courts took Dollond's to be so instead is, in Ramsden's account, an unfortunate piece of legal chicanery.

But the only evidence that Ramsden could point to was not intellectual at all. Hall left nothing that might enable an individual to lay claim to a piece of intellectual property in the eighteenth century: no letters, papers, patents, pamphlets, or books. All that remained were two pieces of shaped glass, not even made by him, from which one was supposed to infer a series of experiments and a theory of dispersion. That Ramsden was prepared to go through such convolutions to try and establish Hall's claim to a philosophical discovery indicates how eager he was to demonstrate his preference for philosophy over trade to Banks and other genteel Fellows of the Royal Society.

Peter Dollond was particularly incensed by this unequal splitting of the credit or, as he preferred to call it, the Truth. Truth, like sovereignty, was indivisible:

> The philosophical Truth of the invention of the achromatic telescope by Mr. Dollond... [should now] appear to be as well established as the legal truth has already been an ingenious distinction—introduced by Mr. Ramsden—a distinction however not very evident to plain minds; as Truth must be the same, whether it be found in Westminster Hall or in the Royal Society.

An essential part of that truth for Peter was that John Dollond had not stolen the idea or design of the achromatic lens from Hall. Peter recalled that in 1758, after the patent became known to the general London optical community, the optical jobber Mr. Bass "recollected to have ground glasses something similar" and wrote to Mr. Hall. Hall then contacted the Dollonds, and "until this time, we had never heard of Mr. Hall." Peter also recollected that when Hall and Bass were subpoenaed in the trial before Lord Mansfield, they were both asked if John Dollond "had got the invention from them," to which they both answered "that they had no reason to think he had." Peter then went on to outline the events of the trials, quoting Mansfield's ruling that the significant commercial sale of the lens, making it generally useful, validated the patent, whereas any prior, secret, artifact did not.[345]

As an outsider with no powerful group of Fellows to support him, Peter Dollond had no means to enforce the publication by the Royal Society of this or his earlier letter. Indeed, even if the President of the Society, Sir Joseph Banks, had been positively inclined towards tradesmen (which he most certainly was not), the Society meticulously avoided associating itself with any controversy involving legal or commercial matters. Peter Dollond did have a patron at the Royal Society, the Astronomer Royal Nevil Maskelyne, who sponsored both of his 1789 letters, but Maskelyne had opposed Banks' election to the Royal Society presidency in 1778 and remained marginalized under Banks' firm rule.[346] Peter never published his account of the Hall story, which remained unexamined for more than two centuries.[347] Ramsden's letter, made public under the clever pseudonym "Veritas" in the *Gentleman's Magazine* the next year, thus remained the final and decisive publication on the matter.[348]

Since then, like the Fellows of the Royal Society in Peter Dollond's day, many historians have proven all too willing to accept without question that an unknown country gentleman like Hall and a pure mathematician like Euler were more likely than John Dollond, shopkeeper, to have invented the achromatic lens and, in a long delayed act of falsification, proved Sir Isaac Newton wrong.

Epilogue

Although Jesse Ramsden was frequently lauded in public as the greatest mathematical and optical instrument maker of his day, in private other Fellows of the Society spoke of him in a deprecating language that was never used for the equally eminent George Graham earlier in the century. Sir Joseph Banks, President of the Society, who praised Ramsden publicly and effusively in his 1795 Copley-Medal address, abused him in private as "illiberal, imprudent and improper."[349] And another customer, George Cholmondeley, who wrote in despair of getting the instrument he wanted, found himself wishing "with all his heart that Ramsden were less able or more tractable."[350]

These complaints did not arise because Ramsden was the kind of unclubbable and suspicious man that John Harrison was. They occurred rather because Ramsden was under immense pressure to satisfy a demand for his instruments, one that he could not always meet. Moreover, in the uneasy years that followed the French Revolution, men like Ramsden—who owned shops, worked with their hands, and increasingly mechanized the production and commercialized the selling of their instruments—had lost some of their social stature, nor could their knowledge be called science. Behind the insatiable demand for Ramsden's instruments and the alarm that other Fellows showed about his social "pretensions," we can detect the rising tempo of industrial production and the changes in social demarcation taking place in Britain from the 1780s onwards.

The twin upheavals of the French Revolution and the intensification of industrialization in Britain brought a growing rigidity to definitions of both gentility and science from the early nineteenth century onwards. By the 1830s, science in London was practiced almost exclusively by "gentlemen" whether in smaller societies like the Geological or larger ones like the British Association for the Advancement of Science. In acknowledgement of the growing correlation of science with gentility, historians of British science in the first two-thirds of the nineteenth century have written books with titles like *Gentleman of science* or subtitles like *Gentlemanly specialists*.[351] In the former work, Jack Morrell and Arnold Thackray identify an intelligentsia or clerisy from which, in the early years of the nineteenth century onwards, came nearly all scientists.

Common to these men was leisure and a secure income from a source that did not require constant attention; those who relied upon industrial, engineering or manufacturing wealth were rare and distanced themselves from the "scenes of their labours."[352] Ramsden was elected to the Royal Society and awarded the Copley Medal primarily for the excellence and abundance of his instruments. Unlike Graham, he did not publish extensively in the Society's journal; nor did he, like Dollond, make a surprising discovery in optical science. However, the exactness of his large mathematical instruments, and of his new machine tool for mass-producing smaller instruments, so astonished his contemporaries that he could not be ignored. His ability to produce extraordinarily accurate large mathematical instruments that made possible further advances in mixed mathematics put him at the pinnacle of a tradition that Graham had begun. Yet at the same time, his extensive machine production of ordinary, smaller, instruments made him, like his brother-in-law Peter Dollond, suspect for being too commercial. Ramsden kept professors, baronets, dukes, and even kings waiting to get his instruments, so overwhelming was the demand for his work. One story, probably apocryphal, claimed he "disdained time restrictions. On one occasion he attended Buckingham House precisely, he supposed, at the time named in the royal mandate. The King remarked that he was punctual as to the day and hour, while late by a whole year."[353] Lesser mortals were kept waiting several years for delivery of his telescopes, mural quadrants, zenith sectors and theodolites. The record length of time to fill an order was probably the 23 years that the Dublin Observatory had to wait for delivery of an altitude and azimuth circle.

Yet buyers throughout Britain and Europe, rather than acquire some other maker's lesser instrument sooner, chose to get a Ramsden instrument later.[354] Most of these clients eventually had their orders filled; as Banks noted in 1795, five years before Ramsden's death:

> Few Observatories were without something that bore his name and no expence was spared by those persons who were able to induce him to undertake the construction of capital instruments, for indeed, there was scarcely one instrument made use of in the whole circle of Astronomical Observations, that had not received eminent improvement from his masterly touch.[355]

Undoubtedly one reason for such delays was his perfectionism and the constant introduction of new features to his instruments. But another reason was that, even with a workshop of 50-or-so men, some of them using machine tools that he had invented, Ramsden simply could not keep up with the demand for his products.

His reputation at home and abroad was huge: to Banks, as we have seen, he was "Promethean" and to the head of the Paris Observatory, Jean-Dominique Cassini, he was a "genius."[356] So great a reputation came from his extensive knowledge of the sciences of optics, mechanics and mathematics; his ability to produce instruments of unprecedented massiveness and accuracy; his dividing engines, which applied industrial techniques to instrument manufacture; and finally his part in the Ordnance Survey of south-east England, which established the distance between the Royal Observatories at Paris and Greenwich. As Banks noted, Ramsden had brought honor to the Royal Society and to England by undertaking the survey that fixed the longitudes of Greenwich and Paris thus increasing "the respect of Foreigners for the national Science of Great Britain," by building the large "capital instruments" of many European national observatories, and by constructing dividing engines that mass-produced accurate sextants.[357]

The huge demand for Ramsden's instruments bred resentments and peculiar behavior, including attempts at industrial espionage, among customers who considered themselves entitled to prompt service.[358] One of Ramsden's most demanding customers was Major-General William Roy who, as a servant of the Board of Ordnance, wanted a theodolite and surveyor's rods to fix the precise distance between the two producers of longitude in Europe, the Royal Observatories at Paris and Greenwich. As had been the case half a century earlier with George Graham, an English instrument maker and Fellow of the Royal Society provided instruments of unprecedented accuracy and delicacy for both the British and the French observers in their endeavor to measure the earth's surface.[359] But there the similarities ended. Ramsden had a more specialized role than Graham and, in what was a complex and expensive job, he made no measurements himself, nor solved any pressing questions in natural philosophy. His contribution came in the form of the instruments that he designed and made for others to use.

The initiative to measure the distance between Paris and London came from France when the Director-General of the Royal Observatory in Paris, Jean-Dominique Cassini (Cassini IV), proposed the project in 1783. The Royal Society quickly took him up on it and procured funding from the government. The main force behind the English effort was Roy, who had first seen the usefulness of accurate maps when he had surveyed the Scottish highlands in order to provide roads for the British army after 1745 as it snuffed out the remnants of the Jacobite rebellion. Cassini came to London in 1787 to check on the progress of the survey. He also came to buy English instruments for both the survey and the Observatory in France, following his complete failure to encourage their manufacture in Paris. As Charles Gillispie notes, he had no option but to accept "the enormous superiority of English over French construction that was the counterpart in manufacturing generally."[360]

Whenever Frenchmen with an interest in science travelled to England in the second half of the eighteenth century, along with the sights of London and its environs—the Tower, the Royal Society, the Society of Arts, Covent Garden, Kew Gardens, Windsor Castle—they also visited the leading instrument makers of London in order to inspect and buy their wares. The men they found in those shops were quite different from those they were accustomed to in France. Unlike the French makers, they were literate and not treated as social inferiors.[361] Faujas de Saint-Fond, on a visit in 1784, considered that English instrument makers enjoyed "a deserved consideration," since a "careful education... the demands of the navy, and the great number of [patrons] of wealth... have concurred to form artists of high regard."[362] However, this "consideration" was soon to recede under the pressures of political and social revolution in France and industrial upheaval in Britain.

In the 1760s, foreigners would visit the mathematical instrument-makers Jonathon and Jeremiah Sisson and John Bird, the clockmakers John Harrison and John Ellicott, and the opticians John and Peter Dollond and James Short; in the 1780s, they visited the mathematical and optical instrument-maker Ramsden, the opticians Peter Dollond and William Herschel, and the philosophical instrument-maker Edward Nairne.[363] While not always impressed with their grasp of advanced mathematics—Lalande thought that Short made his parabolic reflecting mirrors by some form of "groping" or "trial and error"—all the French visitors admitted that the English instrument makers were superior to any others in Europe; they bought their instruments and took them back to France.[364] Thus Cassini was not exceptional in going to London to buy instruments for the Paris Observatory.

However, Cassini did have one extraordinary purpose in mind, namely to bring two workers with him to learn how to make instruments in Ramsden's own workshop and thus to "help" Ramsden meet Cassini's order on time. Cassini laid on the flattery pretty thickly: only Ramsden himself, he wrote, could give the French workers "the precious lessons of a great master." Ramsden politely declined the honor of allowing competitors into his workshop to copy his methods, noting that he already had 50-or-so workers in his shop, sufficient to meet any order, and that he did not wish to incite jealousy among his workmen or disrupt the flow of his business by admitting strangers among them. Cassini hoped that other English makers might prove more pliable.[365] He further proposed paying one French workman the very large sum of 4,000 *livres* (£200)—four times the annual income of a prosperous artisan—to ensure that he would return to France with the methods he had learned in London; but no more was heard of these deliberate attempts at industrial espionage.

Cassini's superior, baron de Breteuil, Secretary of State of the King's Household (Secrétaire d'État à la Maison du Roi), also had enthusiasm for Ramsden's abilities but was not so sanguine as to his punctuality. He wrote to the French ambassador in London to check up on Ramsden, and discovered from him that Ramsden was such a perfectionist that he often had trouble finishing the jobs he had begun.[366] Nevertheless Breteuil conceded that Cassini had no choice but to order a large meridian telescope and a mural quadrant of eight-feet radius from Ramsden.[367] Throughout the entire negotiations, the French referred to Ramsden with the genteel title "le sieur" and treated him as an equal, not as a servant. However, in a sign of the renaissance of French instrument making that would come in the nineteenth century, they did not turn to Ramsden for a surveying instrument. Instead they looked home to the designs of the military engineer Borda whose reflecting circle was executed by the instrument-maker Lenoir.[368] The reflecting circle was compact, light, and clever in design, but not quite as accurate as the massive theodolite that Ramsden was to build for the English portion of the survey.[369]

Ramsden took over three years to complete the new theodolite: too long, as we saw in Chapter Five, for the organizer of the survey in England, General Roy. Roy publicly attacked Ramsden before the Royal Society, claiming that he had not employed enough workmen from the beginning to finish the job in time.[370] This dilatoriness had, Roy asserted, led to "repeated and tedious disappointments." The massiveness, intricacy and accuracy of the theodolite itself, however, remained beyond criticism. Roy grudgingly admitted that "the merits of the whole operation rest in a great measure on the superior excellence of this Instrument."[371] Ramsden, who was present as Roy enumerated his criticisms, replied in a furious letter to the Royal Society protesting the "public abuse" in front of the Fellows and even, indirectly, before "his Majesty, to whom a copy of the above paper was presented." His principal defence was, as Banks noted, that he could not possibly have worked any faster in building a theodolite with so many innovations.[372] Ramsden's letter was not read before the Society, possibly because Roy was in poor health (he died later that year), but also because Banks, who at that point was not talking to Ramsden, would not have brooked such an ungentlemanly outburst at the Society. Instead, he wished to keep the mushrooming and deeply embarrassing dispute as quiet and private as possible.

Roy was not the only Englishman who had to put up with Ramsden's delays in meeting orders for a superior instrument. Another was the Reverend Thomas Hornsby, astronomical tutor to the Duke at Blenheim and, like Bradley earlier in the century, Savilian Professor of Astronomy at Oxford. Hornsby never spoke of Ramsden with the affection and respect with which Bradley spoke of Graham. He continually bemoaned the difficulty in getting Ramsden to deliver instruments on time. While he acknowledged the greatness of Ramsden's instruments, boasting that he had finally acquired for the Duke a six-foot quadrant that was "the most compleat Instrument in the World," he also warned that "if you receive your Instrument in a few Months you may congratulate yourself on your good luck."[373] When writing to the Duke of "the great and immortal Mr. Ramsden" who has given his "Promise to the Winds," he had to beg for ducal intervention with that "intolerable villain that lives near St. James's Church" in Piccadilly.[374] Hornsby certainly knew his place in the scheme of things—on one occasion he congratulated his patron, the Duke, for the arrival of spring—but his letters display a definite resentment towards Ramsden for what he felt was insufficient respect paid to those who ought to be Ramsden's social superiors. Behind Hornsby's mocking of Ramsden as "the Gentleman of Piccadilly" lurks the fear that if men like Ramsden are not put firmly in their place social chaos will follow.[375] Indeed, as Hornsby complained of Ramsden's insufficient obsequiousness in 1791, actual and incontrovertible social chaos was taking place in France.

There is also something sinister about Hornsby's labelling of Ramsden, in 1791, in the language of the French enemy, as "le

Sieur Ramsden." Not only is Ramsden getting too big for his boots, Hornsby implies, but he is of the same ilk as those revolutionaries across the Channel who were destroying an aristocratic and deferential society; the kind of society that, in Britain, the likes of the Duke of Marlborough ruled over. That, on the eve of the French Revolution, foreigners like Cassini and Breteuil had actually and unironically called Ramsden "le Sieur" only made matters worse. By the end of the eighteenth century, social and economic pressures undermined the ability of men like Ramsden to be measurers and experimenters as well as makers, and they became increasingly rare at the Royal Society.

However, I do not wish to close this book on such a bathetic note. By a sensitive application of mixed mathematics and experimental philosophy, instrument makers learned to nullify the variable properties of the materials they worked with and to approach the geometrical ideal. The traditions that Graham and Dollond helped establish—of being observant, sticking to the facts, demonstrating mechanical ingenuity, and above all approaching perfection through the pursuit of exactitude—still lived on in Ramsden. The quest for the greatest humanly possible accuracy had the support of Sir Isaac Newton. Eighteenth-century instrument makers at the Royal Society took him at his word and became "the most perfect mechanics of all."

Appendix A: Contents of the *Philosophical Transactions of the Royal Society of London*

ANA anatomy and physiology of animals and plants

AST astronomy

ATQ antiquities

GEO geography, cartography, navigation, surveying, shape of the earth, astronomical observations to fix latitude and longitude

MEC machines, mechanics, ballistics

MSC miscellaneous including: "agr" (agriculture), "ann" (calculations of annuities, life tables, populations, chances), "hydr" (hydrostatics, hydraulics), "log" (logarithms, calculating and conversion tables), "mus" (music), "pneu" (pneumatics), "opt" (optics), "wm" (weights and measures), "oth" (other)

NH natural histories of: animals, plants, minerals, waters, the atmosphere, the heavens, manufacturing processes, the earth (including topography), nations and peoples

MM mixed mathematics, including: astronomy, geography, mechanics, optics, calculations of annuities, hydrostatics and hydraulics, weights and measures, pneumatics

MED medicine, surgery, pharmacy, bills of mortality

PM pure mathematics

	NH	NP	PM	MM	ANA	MED	ATQ	MSC	
1720	2	5	1	5	2	10	—	agr*3	28
1721	9	4	1	5	5	2	—	agr*2	29
1722	4	3	3	9	7	13	—	agr	39
1723	12	5	2	5	7	14	1	oth*2	48
1724	11	9	—	8	5	10	—	agr	44
1725	8	8	—	7	2	4	—	—	29
1726	10	6	2	12	4	3	—	—	37
1727	12	4	1	7	1	9	—	agr*2	36
1728	10	2	1	12	5	6	1	agr*2	39
1729	8	3	1	10	2	5	—	—	29
Total	86	49	12	80	40	76	2	13	358
%	24	14	3	22	11	21	1	4	100

Table 7: Contents of *Phil. Trans.*, 1720–1729

	NH	NP	PM	MM	ANA	MED	ATQ	MSC	
1730	8	3	—	12	10	2	—	—	34
1731	12	1	1	8	3	7	1	agr	35
1732	11	6	—	7	3	7	—	—	34
1733	11	5	—	11	3	5	—	—	35
1734	13	7	—	6	1	2	—	—	29
1735	7	5	2	7	—	4	4	—	29
1736	12	7	1	16	7	5	2	—	50
1737	7	1	1	11	1	2	1	—	24
1738	14	7	1	10	1	19	1		53
1739	14	13	—	6	—	7	—	—	40
Total	109	55	6	94	29	60	9	1	363
%	30	15	2	26	8	17	2	—	100

Table 8: Contents of *Phil. Trans.*, 1730–1739

	NH	NP	PM	MM	ANA	MED	ATQ	MSC	
1740	14	2	1	6	5	8	8	—	44
1741	17	10	2	13	7	24	3	—	76
1742	18	3	1	11	4	11	1	agr	50
1743	7	2	2	10	10	11	1	—	43
1744	12	3	—	5	7	11	5	agr*3	46
1745	14	11	—	3	5	16	6	agr*5	60
1746	11	15	—	8	7	10	1	agr*4	56
1747	27	9	1	8	4	10	4	—	63
1748	25	11	1	7	4	10	6	oth*3 agr	68
1749	17	3	—	14	2	12	3		51
Total	162	69	8	85	55	123	38	18	557
%	29	12	1	15	10	22	7	3	99

Table 9: Contents of *Phil. Trans.*, 1740–1749

	NH	NP	PM	MM	ANA	MED	ATQ	MSC	
1750	81	8	1	12	3	8	3	—	116
1751	17	7	1	7	4	12	3	—	51
1752	14	7	1	8	1	15	—	—	46
1753	21	9	2	13	—	8	4	—	57
1754	13	11	1	9	2	6	7	agr	50
1755	23	6	1	8	3	12	5	—	58
1756	39	4	—	2	2	5	5	agr	58
1757	22	4	—	12	3	12	5	—	58
1758	24	4	3	7	4	12	3	agr	58
1759	14	8	1	4	2	7	3	—	39
Total	268	68	11	82	24	97	38	3	591
%	45	12	2	14	4	16	6	1	100

Table 10: Contents of *Phil. Trans.*, 1750–1759

	NH	NP	PM	MM	ANA	MED	ATQ	MSC	
1760	15	7	1	13	1	7	3	—	47
1761	17	6	—	29	1	6	2	agr*2	63
1762	15	6	—	14	—	9	1	—	45
1763	23	4	2	15	1	5	4	agr	55
1764	19	2	1	24	2	5	4	—	57
1765	10	4	1	9	2	8	—	agr	35
1766	14	4	—	12	3	5	3	—	41
1767	17	13	2	6	2	7	3	—	50
1768	14	4	2	15	1	4	5	agr*2	47
1769	18	5	—	35	4	3	2	—	67
Total	162	55	9	172	17	59	27	6	507
%	32	11	2	34	3	12	5	1	100

Table 11: Contents of *Phil. Trans.*, 1760–1769

	NH	NP	PM	MM	ANA	MED	ATQ	MSC	
1770	21	6	1	17	4	2	1	oth	53
1771	25	5	1	12	1	2	5	agr*5	56
1772	15	4	1	8	3	2	2	—	35
1773	24	5	1	5	4	—	1	—	40
1774	17	7	—	12	4	7	1	—	48
1775	18	11	4	8	5	3	—	—	49
1776	20	8	3	6	2	5	—	—	44
1777	12	4	2	9	2	4	1	agr*2 oth	37
1778	16	12	4	9	1	7	—	agr	50
1779	15	7	2	8	2	4	—	—	38
Total	183	69	19	94	28	36	11	10	450
%	41	15	4	21	6	8	2	2	99

Table 12: Contents of *Phil. Trans.*, 1770–1779

	AST	GEO	MEC	MSC	
1720	2	2	—	wm	5
1721	3	1	1	—	5
1722	3	2	2	hydr*2	9
1723	2	2	—	opt	5
1724	3	5	—	–	8
1725	3	4	–	—	7
1726	1	10	1	—	12
1727	3	3	1	—	7
1728	3	9	—	—	12
1729	4	5	1	—	10
Total	32	38	6	4	80

Table 13: Mixed Mathematics in *Phil. Trans.*, 1720–1729

	AST	GEO	MEC	MSC	
1730	5	2	4	hydr	12
1731	—	5	3	—	8
1732	1	4	2	–	7
1733	4	7	—	—	11
1734	1	4	1	~	6
1735	—	3	4	—	7
1736	2	7	2	opt*5	16
1737	3	7	1	—	11
1738	4	5	—	hydr	10
1739	3	1	2	—	6
Total	23	45	19	7	94

Table 14: Mixed Mathematics in *Phil. Trans.*, 1730–1739

	AST	GEO	MEC	MSC	
1740	—	3	—	opt*2 wm	6
1741	10	1	—	hydr opt	13
1742	2	6	2	wm	11
1743	7	—	2	wm	10
1744	3	—	1	ann	5
1745	—	—	2	hydr	3
1746	2	1	4	hydr	8
1747	6	1	1	–	8
1748	4	2	—	log	7
1749	6	5	2	wm	14
Total	40	19	14	12	85

Table 15: Mixed Mathematics in *Phil. Trans.*, 1740–1749

	AST	GEO	MEC	MSC	
1750	7	4	1	—	12
1751	5	1	1	—	7
1752	1	—	5	pneu*2	8
1753	4	3	—	log opt*5	13
1754	3	3	1	ann opt	9
1755	4	1	2	ann	8
1756	—	1	—	ann	2
1757	2	3	2	ann*2 hydr pneu*2	12
1758	3	2	—	hydr pneu	7
1759	1	—	3	—	4
Total	30	18	15	19	82

Table 16: Mixed Mathematics in *Phil. Trans.*, 1750–1759

	AST	GEO	MEC	MSC	
1760	8	2	—	mus opt wm	13
1761	23	4	—	ann opt	29
1762	7	2	4	hydr	14
1763	7	3	2	ann opt*2	15
1764	18	3	1	ann hydr	24
1765	2	2	1	hydr log opt*2	9
1766	8	3	—	opt	12
1767	4	1	1	—	6
1768	7	7	1	—	15
1769	22	9	1	ann opt*2	35
Total	106	36	11	19	172

Table 17: Mixed Mathematics in *Phil. Trans.*, 1760–1769

	AST	GEO	MEC	MSC	
1770	8	6	1	ann*2	17
1771	7	4	—	log	12
1772	3	5	—	—	8
1773	2	1	1	ann	5
1774	4	7	—	ann	12
1775	1	1	—	ann*2 hydr mus*2 wm	8
1776	1	2	1	ann*2	6
1777	1	3	2	opt*3	9
1778	3	4	2	—	9
1779	3	2	1	opt*2	8
Total	33	35	8	18	94

Table 18: Mixed Mathematics in *Phil. Trans.*, 1770–1779

Appendix B: Fourth Dialogue by Lorenzo Selva

"Fourth Dialogue," MS Blundell 6. Partial manuscript reproduced with the permission of the Museum of the History of Science, Oxford.

"The inconstancy of the Newtonian Mean of the more refrangible rays to the less, was first asserted by Euler, confirmed by Klingesterna [sic] and afterwards by Dollond."

Prof. The experimental demonstrations of Newton and the declared impossibility of ever being able to arrive at perfection in dioptric telescopes, occasioned, that for an intire age the whole literary world abandoned every attempt towards it and applied all their study to improve the catadioptric. Newton supposed that the mean refraction of the violet rays to the red was constant in every substance, and that it was as 20 to 27. M. Euler of Berlin began the first to doubt thereof and to prove not by experiments but by the analogy of calculations, that it was possible to form object glasses, to inclose water between them and to correct the different refrangibility. He gave at the same time certain figures and combinations for the required sphericity. In vain however did the world attempt to produce the wished for effect.

Revd. I am surprized that Newton who had supposed that an object-glass consisting of two glasses with water enclosed could regulate the aberration of the spherical figure, had not thought of trying all the possible kinds of Crystals, in order to assure himself fundamentally with respect to the erroneously-supposed constant mean refraction if he had but doubted a little upon the subject, he would certainly have prevented Euler.

Prof. It might have been so. But he was far from entertaining that suspicion and Dollond of England opposes Euler, maintaining that the analogy of his calculations could never have force against the experiments of Newton, which determined the aforesaid Mean to be constant in every substance. Euler replied that if the different refraction had the same Mean in every different substance, that our eye consisting of a diversity of humours would not give its focus in one point, and that in all men the sight would be greatly confused. But as the contrary is the case with those who have no defects, we must necessarily conclude, that the eye is the most perfect dioptric machine, and that in it the different substances and perhaps also the figures of the humours correct the errors of the refraction, dispersion and aberration of the rays even by the spherical figure.

Revd. In fact, if all the corrections were not necessary in the eye, its composition might be much more simple, but we see that the omnipotent hand of the creator in its construction has thought a multiplicity of humours and a diversity of figures to be necessary to perfect vision, which might unite, without any dispersion in the different refractions, all the focuses of the heterogeneous rays in one sole point. The same Newton made also of himself this observation and explained it in his Optics; but he never thought of applying himself to imitate Nature, that great mistress, in order to study her seriously in the most arduous undertakings.

Prof. In the meanwhile Klingestema [sic] mathematical professor in the university of Upsal [sic] informed of this literary dispute made some experiments and reversed all at once those of Newton finding that the given Mean of the different refraction passed by him thro' diaphanous bodies and fluids was not the same in all substances, and he recommended the most attentive precautions to be used in making such experiments. Dollond applied himself thereto and by repeated observation found both the laws of Newton as well as those of Euler were false and that the mean refraction in different substances were inconstant and he succeeded in destroying the colours by the use of two prisms, the one of glass and the other of water. With this he began at first to improve telescopes with object-glasses consisting of glass and of water; but these not suceeding because the refractive power of the water being little different from that of the glass, it was necessary to give to the glass a too-convex spherical figure, which consequently too much increased the error arising therefrom; he sought from different crystals the difference of their dispersive power and had the fortune to meet with two kinds of Flint-Glass and of Crown Glass, which possessed a dispersive power different from one another and which was as 3 to 2 whilst in general it is found in common glass to be equal. Having discovered these two different kinds of crystal, he thought, that if he united in the same object-glass one convex-glass of the Crown, which converging the rays would refract and separate them as 2 and another concave glass of the Flint, which diverging the rays would refract and separate them as 3, he should be able to form an object-glass capable of uniting in one single point the more refractive violet with the less refractive red, and that therefore with such an object glass he should remove the colours from telescopes and give them the greatest distinctness. He proved the effect and having succeeded, gave to the public his perfect telescopes. The first which I had in my hand was shewn me by the Abbe Boscovitch then a Jesuit and the same gentleman whom I have cited in another conference. It was one of three feet in length and had an aperture of one inch and and half; for they were not yet very perfect. I was quite astonished indeed at its admirable effect, considering it at first only as a common dioptric telescope; but the Abbe having soon let me see that an object-glass consisting of two different kinds of crystal was the cause—I applied myself immediately to study seriously... and came to understand how the possibility of the correction by the union of two lenses of different materials. ...I procured some Flint glass... twenty small pieces of which however I found only seven to be of service to me; all the rest... were full of twistings and threads, whence they were quite useless... nor was I then in circumstance to lose myself after it, by searching after a composition (matter) believed to be a secret and a mystery. ...A year after, his E[xcellency] the Count de Baschi formerly ambassador of France to this most Serene Republic, my patron, sent me from Paris a proclamation published in 1766 and 1769 which I yet preserve, the same was prolonged with augmentation of rewards unto 1773. In the same was also proposed by the Royal Academy of Sciences a generous reward by the weight of the same King; To whoever knew best to imitate and to succeed best in an exact composition of the crystal called English Flint-glass. [which has at least the same diffractive power, and which does not have striations, filiaments, or threads, and which has no bubbles or point defects]... I applied therefore to examine in the first place physically the composition (matter) by repeated experiments and by dissolving by fire those pieces which I had from England—[after] a thousand repeated experiments. ... [I made glass] which gave the greatest dispersion, and which therefore ought to exceed the English Flint-glass; but I found a considerable quantity of twisting which increased in proportion, and I then conceived that the same thing must have impeded Dollond, being so great that in 20 pieces of crystal come to me from London I was not able to find more than 7 fit to be used. This will have hindered since then the perfection of his Telescope.

Revd. Be assured that to this very day there is great scarcity of good flint-glass in England, and they are studying hard how to purify it and render it free from the threads and particularly from those different strata, which are found in the coarseness of the mass and which palpably hinders the exact refraction of the rays. This difficulty in procuring good Flint in England is confirmed by the Abbe Boscovich in his works and writings. So that a reward is there offered of a 1,000 guineas to whoever will best succeed in it.

Prof. Cospetto! What! more that 2,000 Zecchins! To say the truth, I did not at that time think of these strata (streaks) which were afterward shewn me by the same Abbe Boscovich. However—I had from the furnace by the force of my study and [Sig. Bertolini's] labour a small vessel of glass, which could stand comparison with that of England, and if we respect the judgment at that time given... by the professors of the University of Padua I excelled it. ... In the meantime with this my first flint-glass I made a telescope of about a foot and an half in length with an aperture of one inch and two thirds; because the small quantity made in this first parcel did not yield me a sufficiently large mass from which I could get a piece of it for a greater diameter... two excellent professors of experimental philosophy and Astronomy [at Padua compared]... it with one of judgment of these men was given in writing in these precise terms; "That after various and diligent proofs made by these men, they were convinced that the Venetian Telescope of Selva both for clearness and for the contour... and for the extent of the field and for the magnifying of the objects particularly with the second eye-tube... not only equalled the English telescope but sensibly excelled it in goodness and more particularly in the aperture of the Object-glass." ... [I was granted] the too-honourable title of Public Optician. ... [the Prof. then explains the meaning of the word "achromatic" to the Count] Fortunately for me the above mentioned Abbe Boscovich then Public Professor of Astronomy in the

University of Milan returned to Venice... honouring me with many visits and conferences and making with me many experiments upon my Flint, in this very room of mine. ...

Revd. In many parts of his works he does you justice, and I will quote you the very words which he writes, speaking of his Variable Vitrometer in his first work Sec.II n.22 "I have," says he, "of common glass, of English flint, of Venetian Flint which has a still greater refractive and dispersive power. This, that is your Venetian, is in my opinion to be preferred to the rest even to the English."

Prof. I am much obliged to him. He then told me, that I had not obtained all in my flint; but also that I should never arrive at any perfection in such telescopes, if I did not learn the mode of calculating the different refrangible and dispersive powers as well of flint-glass as of common crystal, which I should wish to unite in order to have the exact proportions of the rays of the spherical surfaces to give them not only to the object-glasses but also the eye-glasses, which ought to be compounded in order to take away compleatly the colours from the Dollondians. In fact he makes to appear in his cited Memoirs, and has assured me of the same, that the colour one sees upon the margin of object glasses, when he uses telescopes, do not come from the object-glass but from the eye-glass, and he writes in Sec.III, n.33 that in all the achromatic telescopes made by Dollond, having an eye-glass of simple Flint, in the field and in the increased size, he had observed at the edge of the same field those colours most sensibly and much stronger and brighter than in common telescopes and that therefore that they could never be called achromatics if the eye-glasses were not corrected. This was known to me ever since I saw, in the appendix by P. Pezenas professor of Hydrography in Marseilles translator of Smith's optics where he treats of achromatic telescopes, a problem, in which is given the manner of making compound eye-glasses to take away the colours; and the Abbe Boscovich afterwards proved evidently in his last work that the eye-glasses in telescopes even with an achromatic object glass occasioned colours, as he experienced in Gregorians and which you have already mentioned; whence he with reason infers, that one can never call a telescope achromatic, unless the colours be also taken away from the eye-glasses. ...First of all it is necessary that the artificer understand the refractive and dispersive power as well of flint as of the common glass which he would unite with it, since that a vessel of flint-glass taken out of the same furnace will have one (power) and another (vessel) a different; thus the same kind of glass will always have in every part a different (power) from another part and one can never join to flint any sort whatever of common crystal, except that glass of which he understands the forementioned power.

Revd. The Abbe Boscovich required that that same eye-glass besides being simple or single, should be of one kind alone of glass. I am surprized, when, in any of Dollond's telescopes which they call good, I see sum [sic] of the two lenses, even sometimes three different kinds of glass: I must needs say that the workman had tried the refractive and dispersive power of each kind and having made his calculations united them together as he could.

Prof. These experiments and calculations cannot be made but by the use of a variable Vitrometer (perhaps unknown to any of those workmen) which was discovered by the inestimable Abbe Boscovich and which I executed under his directions, as you see in this one which demonstrates it. I have made many others besides both for him and for his Ex. Count de Durazza formerly the imperial Ambassador and for many other amateurs of optics, as well as Mathematical professors.

Revd. He himself (Boscovich) says in the preface of the first book; Its form came into my mind at Venice on 1773 where I took care to have it made by that excellent artist of Telescopes Dominic

Selva (calling you by the name of your father who was dead many years before) many others in imitations thereof were then made both these (i.e. by you, because I know that you are their sole maker) and in other places.

Prof. Neither flint, nor vitrometers are to this day made by any one else but me. I think it is necessary to shew the Count in some degree the use of it. ...

Prof. ...It happened to me some years ago that a respectable gentleman took it into his head that I should be able to make him a perfect achromatic telescope like to one of Dollond's which had come expressly from London to his Ex[cellency]. the Count Firmien, confining me exactly to the same dimensions of the rays taken most exactly, to the same size, both of the aperture not only of the two lenses which compose the object-glass and of its five eye-glasses with their distances; but also of the Diaphragm, (division) and of the very tubes which form the telescope. Nor was it possible to persuade him, that in defiance of the exactness of my work it would be impossible to give him a telescope of equal perfection, if I had not the same kinds of flint-glass as also of the other glasses which were of several kinds; as was made plain to me in the different colouring of the crystal of the eye-glasses described to me: persuading him that doubtless every kind of flint and of glass has a different refractive and dispersive power which require a different measure in the rays of the spherical surfaces in order to destroy their colours and to render a Dollondian telescope perfect. In short I was successful in conquering him and I made him quite content by giving him one of mine of different dimensions shorter by two inches than the three-feet one prescribed. ...

Revd. The Abbe Boscovich in his extract from the first Tract Section XV gives the result of the different strength (valori) thro' the different kinds of glass, and infers therefrom the necessity of determining the quality of those which one would use, with the impossibility to give general rules to workmen (if chemistry does not discover glass of a constant quality) and the error those men labour under who imitate the measures found in good telescopes because used upon unknown glasses. ...

Count It appears to me indeed to be a thing of great embarrassment and a loss of time to be obliged as occasion requires, to reduce by the aforementioned methods every convex and concave superficies of every object glass or compound lens to such very small proportionate parts; and I admire your patience.

Prof. But this is indispensible; tho' to some people the price of a Dollondian telescope seems immense, because they reflect not upon the loss of time, which such combinations and calculations occasion; besides the labour which is necessary to work and polish exactly all the lenses; tho' I have for some time taken upon me this change to prepare them and polish them upon papers for my sons. ...

Revd. ... Abbe Boscovich in the 2nd book of his works... proves that the greatest error of the object-glass which occasions the greatest confusion in the image is that of the spherical figure. He directs to take that away by the union of two lenses of the same kind of glass and to arrange the 4 eye-glass in such an order as to correct the errors of the sphericity and to destroy the colours. ... He does not restrict the abilities of the workmen to them [the rules] and allows that practice does more sometimes than theory (grammar); so much that after he has the measures of the focal distances and of the apertures, he concludes; "Experience and trials will determine these things better than theory, which is too complicated often."

Abbreviations

DNB	*Dictionary of National Biography*, Oxford University Press, Oxford, 1920-1921.
DSB	*Dictionary of Scientific Biography*, ed. C. C. Gillispie, Scribner, New York, 1970-1980.
Experiment VIII	Exp. VIII, Bk. 1, Pt. ii, Prop. 3, *Opticks*, Isaac Newton, London, 1704.
F.R.S.	Fellow of the Royal Society of London
Mémoires	*Mémoires de L'Académie Royale des Sciences, Paris*
MS Blundell 6	"Fourth Dialogue" (a partial translation of Selva 1787), MS Blundell 6, Museum of the History of Science, Oxford. The Blundell Manuscripts are the papers of the Dollond family and MS 6 dates from 1787 to 1789 and was made for Peter Dollond.
Britannica	*Encyclopaedia Britannica*, 11th Edition, Edinburgh, 1911.
Phil. Trans.	*Philosophical Transactions of the Royal Society of London*

PRO	Public Record Office, London. All references are to records held in the Chancery Lane archive unless otherwise indicated. The following prefixes correspond to the following courts or administrative bodies: C - Chancery; K or KB - King's Bench; PC - Privy Council; SO - Signet Office.
RASR	Royal Astronomical Society, London, Radcliffe papers.
RGO	Royal Greenwich Observatory, London.
RSCMB	Royal Society Council Minute Book (copy unless otherwise indicated). All Royal Society archive references are to the Society's Library at 6 Carlton House Terrace, London.
RSEL.x	Royal Society Early Letters; 'x' denotes the first letter of the writer's name.
RSJB	Royal Society Journal Book (minutes of the weekly meetings; copy unless otherwise indicated).
RSL&P	Royal Society Letters and Papers.
RSLB	Royal Society Letter Book (accounts of letters received).
RSMM	Royal Society Miscellaneous Manuscripts.
SLSC	Sutro Library, San Francisco, Banks Correspondence, SC.

Endnotes

[1] Newton 1999: 381.

[2] Entries for Dec. 24, 31, 1675 in Hooke 1935a.

[3] Quoted in Nelthorp 1873: 97, who states that this inscription was "unfortunately... removed in the year 1838."

[4] Warner 1990, Bud & Warner 1998, Taub 2009.

[5] Morton & Wess 1993, Golinski 1992, Schaffer 1983, Stewart 1992, Walters 1997.

[6] See Ch. 8, "Invisible technicians" in Shapin 1994.

[7] Bennett 2011, Schaffer 2011, Taub 2011, Baker 2010, Baker 2009, Jacob 1997, McConnell 1994, Chapman 1990, Anthony Turner 1987, Bennett 1987, Robischon 1983, Heilbron 1982, Howse 1975, Cardwell 1972, Musson & Robinson 1969, Taylor 1966, Daumas 1953.

[8] Heilbron 1982: 188, Cardwell 1972: 17-18, Jacob 1997: 114-115.

[9] McConnell 1992, Millburn 2000, and A. D. Morrison-Low 2007 have made excellent use of the archives that have survived.

[10] Cochrane 1956.

[11] van Helden 1983: 64.

[12] Aphorism 95, Bk. 1, in Bacon 1960.

[13] Galison 1997.

[14] Aphorism 94, Bk. 1, in Bacon 1960.

[15] For the problems the Society faced in extracting trade secrets from tradesmen see Boas Hall 1983.

[16] Sobel 1995.

[17] Reynolds 1797: 127.

[18] Copley Medal presidential address, RSJB, Nov. 30, 1795.

[19] Sorrenson 1995.

[20] Kuhn 1976, Heilbron 1993.

[21] van Helden 1983, Jardine 1999.

[22] Mahoney 1996: 66.

[23] Sprat 1667: Section 21.

[24] van Helden 1983: 67.

[25] Heilbron 1993: 104.

[26] John Harrison's grandson parleyed the Longitude Prize into a Lord Lieutenantship in Wales. See the entry under "Harrison, John" in *DNB*.

[27] Ginn 1991.

[28] Sorrenson 1999, Sorrenson 2001.

[29] Bennett 1985b.

[30] Hamilton 1926, Macleod 1987, McKendrick 1982.

[31] Brewer 1988, Howse 1975, Gerard Turner 1987, Hankins & Silverman 1995.

[32] Bennett 1986.

[33] Price 1984: 54.

[34] Price 1980: 76.

[35] Brown 1979: 5, 19, Robischon 1983: 334.

[36] Campbell 1747: 306.

[37] Brown 1979: 5.

[38] Campbell 1747: 253-254, 336. Campbell estimated that it cost £20 to £50 to apprentice a boy to the trade and between £100 and £1000 to set up as a Master.

[39] Campbell 1747: 250-252.

[40] Chapman 1990: 81, and more generally Morrison-Low 2007.

[41] For the exclusion of instrument makers from the Royal Society in the nineteenth century see Bennett 1985a and the excellent but alas unpublished Ginn 1991.

[42] Sorrenson 1993: 7.

[43] Wilson 1799: 169.

[44] Sorrenson 1995.

[45] Golinski 1999.

[46] See their individual entries in Clifton 1995.

[47] Taylor 1966, Clifton 1995.

[48] Morrison-Low 2007.

[49] Bennett 1985a: 24 and Ginn 1991: vi respectively.

[50] Symonds 1969.

[51] The extant diary which ends on Aug. 8, 1693, is not continuous; 1681-1688, 1690-1692 are missing. Hooke varied the spelling of Tompion's name in the diary.

[52] Hooke 1935a. Garaways was a coffeehouse in Change Alley Cornhill. More oversaw the building of the Royal Greenwich Observatory and Flamsteed was its first incumbent; Tompion was subsequently to work for both of them.

[53] Foster 1659, Streete 1661.

[54] Entry for Apr. 27, 1675/76 in Hooke 1935a.

[55] Entry for Friday Mar. 24, 1675/76 in Hooke 1935a. Mans was a coffeehouse in Chancery Lane.

[56] Entry for Sunday Oct. 4, 1674 in Hooke 1935a.

[57] Entry for Sunday Apr. 9, 1676 in Hooke 1935a.

[58] Hooke 1674: 90.

[59] Flamsteed 1835: 45.

[60] Howse 1975: 21-24, 75, 125-131.

[61] Hooke 1674: 78.

[62] Westfall 1980: 250.

[63] RSCM, Apr. 27, 1730.

[64] RSCM, May 9, 1730. He was paid £32 for the cost of materials and subcontracted parts.

[65] Boas Hall 1984.

[66] For some trenchant criticisms of the former see Rousseau & Haycock 1999.

[67] Klein 1995: 363.

[68] Wahrman 1995: 37-39, 55.

[69] "Of refinement in the Arts," first published in 1752, in Hume 1989: 270-271.

[70] Hume 1989: 271.

[71] Hume 1989: 273.

[72] Hessenbruch 1999: 212-215, Sorrenson 1995.

[73] Wise 1993: 247.

[74] Solkin 1993: 131.

[75] Babbage 1830: 1-2, xiii.

[76] Weld 1848, 1858; Lyons 1944.

[77] Copley Medal presidential address, Nov. 30, 1795, RSJB.

[78] Henry Brougham, quoting Sir Joseph Banks, in Weld 1858: 163.

[79] Weld 1858: 43, Bluhm 1958.

[80] Brougham quoting Banks in Weld 1858: 163.

[81] RSJB, Nov 30, 1748.

[82] RSJB, June 24, 1725; April 28, 1726.

[83] RSJB, Dec 15, 1743; April 26, 1750. See the entries under "Patrick, John" and "Bird, John," in Clifton 1995.

[84] RSJB, June 8, 1727; Lord Macclesfield quoted in fn. 17 in Weld 1848: 387. For the first few years the income from the bequest was given to the Curator (Desaguliers), see Weld 1848: 385; see also Bektas & Crosland 1992.

[85] RSJB, Nov. 30, 1748.

[86] RSJB, Nov. 30, 1749.

[87] RSJB Nov. 30: 1758 (*re* Dollond); 1757 (*re* Cavendish); 1759 (*re* Smeaton).

[88] Jacob 1988: 142-151.

[89] Desaguliers 1734, Preface

[90] RSJB, Mar. 22, 1733.

[91] RSJB, Feb. 1, 1739.

[92] Heilbron 1982, Stewart 1992.

[93] RSJB, May 13, 1731.

[94] James Ferguson to the Society, RSJB, Jan. 31, 1744/5.

[95] RSJB, May 22, 1740.

[96] RSJB, Dec. 13, 1753.

[97] Newton 1999: 796.

[98] RSJB, June 12, 1729; June 18, 1747; Mar. 29, 1753.

[99] RSJB May 12, 1759.

[100] RSJB, Nov. 10, 1757.

[101] RSJB, May 14, 1730; Nov. 30, 1759; June 8, 1758.

[102] RSJB, Apr. 24, 1740.

[103] See Pt. II, Sect. xvii, "Judging the Matter of Fact," in Sprat 1667, and Ch. 2, "The Experimental Production of Pneumatic Facts," in Shapin & Schaffer 1985.

[104] RSJB, Feb. 18, 1741/2.

[105] RSJB, Dec. 17, 1747.

[106] Hankins 1985: 40.

[107] RSJB, Dec. 13, 1722.

[108] RSJB, Dec. 10, 1724.

[109] RSJB, Dec. 17, 1724.

[110] Stigler 1986: 44, 54.

[111] RSJB, Jan. 18, 1753.

[112] Schaffer 1988: 116, 124.

[113] Biagioli 1996: 209.

[114] RSJB, Nov. 30, 1753.

[115] RSCMB, Nov. 6, 1729.

[116] Weld 1858: 87.

[117] Hunter 1981: 42-50.

[118] RSCMB, July 12, 1742.

[119] Sprat 1667: Pt. 2, Sect. xxi.

[120] Bacon 1960: Bk. 1, Aphorism 84, Sprat 1667: Pt. 1, Sect. xxi.

[121] RSCMB, Feb. 15, 20, 27; Mar. 12, 26; May 28; Nov. 17, 1752. A life membership in the Society was raised from 21 guineas to 26 guineas in 1766, RSCMB, 8 Dec, 1766.

[122] RSCMB, May 10, 1753.

[123] Sorrenson 1996a.

[124] Walters 1997.

[125] Cochrane 1956: 137.

[126] RSJB, Oct. 26, 1738; Feb. 3, 1732; Jan. 26, 1758.

[127] See entry under "Physiology," in Harris 1710.

[128] Experiments (*Chem* in Table 4) are pretty equally divided between chemical and electrical subjects.

[129] Bazerman 1988: 63. He sampled volumes 1, 5, 10, 15, 20, 25, 30, 35, 40, 50, 60, 70, 80, & 90.

[130] Boas Hall 1991: 139.

[131] Forster 1757: 461.

[132] See Da Costa 2009 for a detailed account of medical and natural historical cusiosities at the eighteenth-century Royal Society.

[133] Daston 1980: 243.

[134] Delisle to John Bevis, RSJB, Mar. 13, 1738/39.

[135] Sutton 1995.

[136] For examples of F.R.S. naturalists who, exceptionally, relied upon their science for a living, see the entries under "Baker, Henry" and "Forster, Johann" in *DSB*.

[137] Feingold 2001, Rusnock 1996: 33-35.

[138] Money 1977: 98.

[139] Hunter 1981: 77.

[140] Quoted in King 1996: 183; Dollond was called this in Kelly 1808: 6; see the entry under "Ramsden," in *DNB*.

[141] Quoted in Millburn 1973: 17.

[142] RSJB, Jan. 19, 1743/44 and Apr. 19, 1744.

[143] See entry under "Cuff, John," in Clifton 1995.

[144] RSJB, Dec. 10, 1730. See also Jungnickel & McCormmach 1999: 57.

[145] The situation was very different for other European scientific institutions. See Biagioli 1996.

[146] Heilbron 1983: 36.

[147] RSJB, May 11, 1727.

[148] RSJB May 28, 1727.

[149] Heilbron 1983: 37-38, Rusnock 1996: 33-35, Feingold 2001.

[150] The following are called "our worthy brother," in the RSJB: Gowin Knight, Nov. 30, 1747; George Graham, Nov. 30, 1748; Benjamin Robins, March 14, 1750/51. It is difficult to tell from the minutes whether it is the Secretary of the Royal Society using this term, or the speaker describing their work.

[151] RSJB, Feb. 13, 1734/35; Mar. 27, 1735. Copies of Wreden's and Gambier's election certificates appear in the RSJB on Nov. 14, 1734. Copies of Locke's and Clare's certificates appear in the RSJB on Jan. 9, 1734/35.

[152] I have begun the table in 1735 since election certificates, which describe the candidates' merits and occupation or social position, were not required until 1731. RSJB, Dec. 10, 1730.

[153] Table 5 in Hunter 1976.

[154] Ch. 2, "Pre-Newtonian and Newtonian science in England," in d'Espinasse 1956.

[155] See Ch. 3, "The Progress of Politeness," in Langford 1989.

[156] Mathias 1979: 158-159.

[157] "Of the different Progress of Opulence in different Nations," in Smith 1976.

[158] Feingold 2001.

[159] Quoted in Heilbron 1993: 86-87.

[160] Heilbron 1993: 90.

[161] Jungnickel & McCormmach 1999: 335-351.

[162] Boas Hall 1984.

[163] Chapman 1990, "Introduction"; Heilbron 1982, "Introduction"; Heilbron 1993: 106-107.

[164] Graham 1722: 198-199.

[165] RSJB, Nov. 14, 1728.

[166] Bradley 1728: 638.

[167] RSJB Jan. 14, 1748.

[168] RSJB, Nov. 14, 1728.

[169] Bradley 1728: 641, 645-653, Bradley 1832: xxvi-xxvii, 202.

[170] Thomson 1812: 346.

[171] The modern value is 8 minutes 19 seconds. See the entry under "Sun" in *Britannica.*

[172] RSJB, Jan. 16, 1729.

[173] RSJB, Nov. 14, 1728.

[174] Edmund Halley, RSJB, Nov. 14, 1728.

[175] Bradley 1748: 2.

[176] Bradley 1748: 6. "Contrive" means here, as it usually did in the seventeenth and eighteenth centuries, to design.

[177] Bennett 1986: 2.

[178] Ch. 4, "The Expedition to Lapland," in the delightful biography of Maupertuis by Terrall 2002, and more generally in Greenberg 1995.

[179] Newton 1999: 821-826.

[180] Iliffe 1993: 365.

[181] Maupertuis 1738a & 1738b.

[182] Newton 1999: 826-832.

[183] Bradley 1734: 303.

[184] Bradley 1734: 306.

[185] Bradley 1734: 304, 306, 310, 311.

[186] Graham 1726.

[187] Chang 2004.

[188] RSJB Jan. 1735/36; Apr. 1, 1736; the eclipse was on Mar. 15, 1735/36.

[189] Newton 1999: 824. The modern value of the average ellipticity of the earth is 1 part in 299. See the entry under "Earth, figure of," in *Britannica*.

[190] The details of the procedure are laid out in Ch. 4 "The Expedition to Lapland" in Terrall 2002.

[191] Maupertuis 1738b: 38.

[192] Terrall 1992: 227-228.

[193] The election certificates of all four explicitly mention the expedition. Celsius was elected as he prepared for the expedition in Jan. 1735/36, Clairaut and Le Monnier immediately after the expedition (Oct. 1737, and Apr. 1739 respectively), and Camus very much later (Jan. 1764).

[194] Celsius to Royal Society, RSJB: May 20, June 10, 1736; Jan. 13, Mar. 24, 1736/37; May 19, 1737. Clairaut to Royal Society, RSJB: Mar. 24, 1736/37; Nov. 3, 1737; Dec. 15, 1737. Maupertuis to Royal Society, RSJB: Oct. 27, 1737.

[195] Bradley translated this letter and sent it to Graham. Maupertuis to Bradley, 27 Sept. 1737, in Bradley 1832: 405. This letter is referred to in RSLB(Copy).24.8, and in RSJB Oct. 27 1737.

[196] Le Monnier to Graham, RSEL.M.3.58, Feb. 1737/38. Other recipients included Bradley, Halley, Stirling, and Desaguliers, RSEL.M.60.

[197] Maupertuis 1738b: 123-128, 99, 154.

[198] Graham to Bradley, London, July 25, 1732, in Bradley 1832: 397.

[199] Beaglehole 1968: ci.

[200] Bougainville 1772: xxvi, La Pérouse 1799: 441.

[201] Sorrenson 1996b.

[202] Woolf 1959.

[203] Beaglehole 1974: 148-149.

[204] Beaglehole 1974:148-149.

[205] David 1988: xxviii.

[206] David 1988: xxix.

[207] Carter 1988: 23.

[208] Beaglehole 1968: xxi-cxxii.

[209] David 1988: xxxiv-xxxvii.

[210] Taken from Hawkesworth 1773. This was the map most familiar to eighteenth-century readers, and it follows very closely the manuscript copies of Cook's own maps.

[211] Cook J1: 274.

[212] Salmond 1991.

[213] Cook J1: 274.

[214] Editorial fn. 2 in Cook J1: 253.

[215] Cook C2: 4-5.

[216] Cook J2: 643.

[217] Plate facing p. 1 in Cook 1777.

[218] Cook J2: 643.

[219] Cook J2: 78-79.

[220] Bennett 1987: 51-55.

[221] See entry under "Magnetism, terrestrial," in *Britannica*.

[222] Fara 1996: 135.

[223] Graham 1724: 96-98, 107, 101.

[224] Good 1998: 352. The modern explanation attributes the phenomenon to the sun's own magnetic activity rather than its heating power upon the earth in Good 1998: 510.

[225] Deiman 1994: 1159-1160.

[226] Shapiro 1979: 126-127.

[227] Euler 1747: 13.

[228] See the entry under "Eye" in *Britannica.*

[229] See entries under "Short, James" and "Dollond, John (I)" in the invaluable Clifton 1995.

[230] Kelly 1808: 6-7.

[231] Stewart & Weindling 1995.

[232] Dollond 1753: 289, 291.

[233] See the entry "Dollond, Peter," in Clifton 1995 and Brown 1979: 8 for a further account of the Spectaclemakers.

[234] "An Account of Mr. Klinginstierna's Paper entitled Considerati circa Legem rafractionis Radiorum luminis diversi generis in mediis diversis. Vid. Newt: Opt., Lib. I. Part II. Prop. 3. Exp. 8." read by James Short at the Royal Society 16 Apr. 1761, RSL&P.IV:66.

[235] Shapiro 1979.

[236] Dollond 1758: 738, 736.

[237] Pipping 1991: 101-103, Daumas 1989: 145. For the attempts to mathematize the discovery, see Hutchison 1991. Thomas Pynchon refers to the excellent qualities of Dollond's telescopes in Pynchon 1997: 13.

[238] Lord Macclesfield, President of the Royal Society, Copley Medal speech, RSJB, Nov. 30, 1758.

[239] Sorrenson & Burnett 1998.

[240] Friedman 1992: 61.

[241] Newton 1999: 381-382.

[242] Friedman 1992: 77.

[243] "The report of the committee... to view the state of the instruments," RGO 4/307.2.

[244] Smith 1738: 332, Maskelyne 1782: i, Ludlam 1786, Howse 1975: 22.

[245] Smith 1738: 332.

[246] Chapman 1990: 67.

[247] An angle cannot be trisected "by means of the straight line and circle" Heath 1981: 235.

[248] See the article "Graduation" in Chambers 1820.

[249] Ramsden 1777, Ramsden 1779.

[250] Nevil Maskelyne, "Preface," in Ramsden 1777.

[251] Sir Joseph Banks, Copley Medal presidential address, RSJB, Nov. 30, 1795.

[252] Ramsden 1779: 3.

[253] Woodbury 1972: 69-71, Steeds 1969: 18-20.

[254] Ramsden 1777. The dividing engine can still be seen in the National Museum of American History of the Smithsonian Institution in Washington D.C.

[255] Ramsden 1777.

[256] Chapman 1990: 131.

[257] Graham 1726: 40-43.

[258] RSJB, Apr. 8, 1736.

[259] Ellicott 1736.

[260] Ellicott 1752.

[261] Ellicott 1752: 485.

[262] RSJB, June 4, 1752.

[263] Smeaton 1754.

[264] RSJB, May 23, 1754.

[265] Pringle 1773: 28.

[266] Horrins 1830: 215. This rather eccentric book, by "Johan Horrins" (anagram for John Harrison), appears to have been written by a grandson of the clockmaker.

[267] Quill 1963: 157-158.

[268] Quill 1963: 152. There is no evidence in the Journal Books, or election certificates that John Harrison ever applied for membership in the Society.

[269] For a detailed and impressive account of Harrison's methods see Burgess 1996.

[270] RSJB, Nov. 30, 1749.

[271] RS election certificates, 1765.

[272] Clark 1999: 444.

[273] Casanovas 1987: 64.

[274] Boscovich 1785: 2. "Adhibui in ipso Dollondiani comperti initio alias methodos..." Comperti could be translated as "experience."

[275] Pl. II, Pt. 1, Vol. 1 of Boscovich 1785.

[276] Boscovich 1785: xxxvi.

[277] Selva 1787.

[278] For an excellent account of the making of flint glass in late seventeenth-century England see MacLeod 1987. See also the entry under "Glass (optical)" in *Britannica.*

[279] MS Blundell 6: 15.

[280] MS Blundell 6: 12-15.

[281] Royal Society of Arts, London, (RSA), Minutes of Committees, March 2, 1771. Russell was awarded £40.

[282] RSMM.14.

[283] MS Blundell 6: 15.

[284] MS Blundell 6: 27.

[285] Priestley 1772: 729, Montucla 1802:447, Herschel 1827: 422.

[286] Dollond 1758: 733. The Earl of Macclesfield, Copley Medal presidential address, RSJB: Nov. 30, 1758; Nov. 27, 1760.

[287] Hutchison 1991, Nordenmark & Nordström 1938 & 1939, Boegehold 1943. See also the extensive and useful bibliography in Fellman 1983.

[288] PRO SO7/233.

[289] Fig. 8, Pl. 4 in Martin 1759.

[290] Willach 1996: 199-205.

[291] Macleod 1988: 76-77; and Ch. 5, "The decision to patent."

[292] Francis Watkins, PRO C12/1956/19.

[293] See the entry under "Ramsden" in *DNB*.

[294] Peter took charge of the business and all aspects of the patent trial. Peter was the "Administrator" of the estate (PRO C33/427) and gives a brief description of the dispute between himself and Watkins in, Peter Dollond to the Lord Chancellor, 12 July, 1765, PRO C12/1956/19.

[295] Its asphericity had been earlier commented on in Kiltz 1942: 41-46.

[296] Willach 1996: 205-208.

[297] PRO C12/1956/19. For Peter Dollond's increased prices after he successfully defended his patent, see Robischon 1983: 292.

[298] Francis Watkins, the Dollonds' former partner, asserted in 1765 that Peter Dollond's profit margin was at least 200%, leading to a minimum annual profit of £800, PRO C12/1956/19.

[299] Macleod 1988: 61.

[300] Macleod 1988: 7.

[301] PRO SO7/233.

[302] Robischon 1983: 300-302.

[303] PRO PC1/7 n.94, received June 22, 1764. The petition, supported by the Spectaclemakers' Company, was signed by 34 instrument makers, mainly optical instrument makers.

[304] The Privy Council no longer regulated patents after 1753, Macleod 1988: 73, 66.

[305] Brown 1979: 20; Thompson 1968:595.

[306] The records of these trials at the London Public Record Office are not complete. The most extensive are a series of depositions dated Oct. and Nov., 1765, in the Chancery Court (PRO C12/1956/19). Peter Dollond seems to have gone to Chancery to enforce an earlier judgment in his favor at King's Bench, July 12, 1764, before Lord Mansfield. The defendants before Mansfield were William Eastland, Francis Watkins, Addison Smith, James Champneys and Christopher Steadman. Dollond was still trying to enforce the judgment in July 1767 (PRO C33/427/296). Dollond won another court case in front of Lord Camden at Common Pleas, in Feb., 1766. The defendant there was James Champneys (*The Gazetteer and New Daily Advertiser*, Feb. 20, 1766).

[307] Francis Watkins, PRO C12/1956/19.

[308] John Bird, witness before Lord Camden, Common Pleas, 1766. This testimony was recalled in Dollond 1789b.

[309] James Champneys, October 29, 1765. Watkins, Smith, Eastland and Steadman use very similar language, in PRO C12/1956/19.

[310] Hall is called this in Nordenmark & Nordström 1939: 380.

[311] Willach 1996: 195.

[312] Christopher Steadman, 1765, PRO C12/1956/19.

[313] Addison Smith, PRO C12/1956/19.

[314] Dollond 1789b.

[315] In another case, before Lord Camden, James Champneys was ordered to pay £150 in damages. *The Gazetteer and New Daily Advertiser*, Feb. 20, 1766.

[316] King 1955: 158-159. Reputation alone could last for a very long time. Van Helden notes that Galileo's reputation as a preeminent telescope maker lasted until his death, even though others had probably surpassed him by 1612, in van Helden 1994: 18. There are hundreds of Dollond Crown Forward Aspherical achromats surviving today; Rolf Willach, personal communication. Dollond & Son probably did not dominate the spyglass portion of the refracting telescope trade after 1772; Deborah Warner, personal communication.

[317] Euler 1768: 220-221.

[318] Euler 1753: 172.

[319] Euler 1768: 221.

[320] Editorial fn. in Vol. 2 of Euler 1823: 350.

[321] Euler 1765: 175; Euler, quoted in Priestley 1772: 470; and Euler, quoted in Dollond 1789a.

[322] Euler 1765: 177.

[323] Hutchison 1991.

[324] Euler, quoted in Dollond 1789a.

[325] Clairaut 1761: 387.

[326] Priestley 1772: 469.

[327] Fuss 1783: 41.

[328] Dollond 1789a, Lalande 1771, Vol. 2: 736, Cassini 1785: 106.

[329] Dollond 1789c.

[330] Ramsden 1789.

[331] 1771 is probably the correct date of Hall's death. See the entry under "Hall, Chester Moor," in *DNB*.

[332] Addison Smith, in a letter to Ramsden, June 4, 1789, quoted in Ramsden 1789. The letter has not survived.

[333] van Helden 1975: 256-259.

[334] Willach 1997: 5.

[335] Francis Watkins, in a letter to Ramsden, c. 1789, quoted in Ramsden 1789.

[336] Dollond 1789b.

[337] Dollond 1789a.

[338] Ramsden 1789.

[339] Dollond 1789b.

[340] Widmalm 1990.

[341] William Roy, RSJB, Feb. 25, 1790.

[342] Jesse Ramsden to the Council of the Royal Society, May 13, 1790, RSDM 4.44.

[343] May 13, 1790, RSMM 3.30.

[344] Sir Joseph Banks, Copley Medal presidential address, RSJB, Nov. 30, 1795.

[345] Dollond 1789b.

[346] Heilbron 1983: 83-91.

[347] Until Sorrenson 1993: 177, Bennett 1998: 4-5.

[348] Ramsden 1790 is very similar to the letter that Ramsden sent to the Royal Society the year before.

[349] Banks to George Rose, Apr. 11, 1793, SLBC, SC 1:50. I thank John Gascoigne for giving me a copy of his transcription of this and other letters from the Sutro Library.

[350] Cholmondeley to Banks, Feb. 2, 1793, SLBC, SC 1:49.

[351] Morrell & Thackray 1981, Rudwick 1985.

[352] Morrell & Thackray 1981: 17. For the increased "distancing" of the English elite from the sources of industrial as opposed to professional wealth, see Wiener 1981.

[353] See entry under "Ramsden, Jesse" in *DNB*.

[354] The instrument was eventually delivered in 1810, ten years after Ramsden's death, King 1955: 169.

[355] Sir Joseph Banks, Copley Medal presidential address, RSJB, Nov. 30, 1795.

[356] Cassini in a letter from London to Breuteuil in 1787, quoted in Wolf 1902: 288.

[357] Copley Medal presidential address, RSJB, Nov. 30, 1795.

[358] For attempts to copy Ramsden's and other London makers' techniques see Christensen 1993. For industrial espionage in general see Harris 1998.

[359] Ramsden was elected F.R.S. in 1786. William Roy was one of his recommenders. Royal Society election certificates, 1786.

[360] Gillispie 1980: 119.

[361] Gillispie 1980: 122.

[362] Saint-Fond 1784: 91.

[363] Ferner to Wargentin, October 1759, quoted in Nordenmark & Nordström 1939: 48. See also Lalande 1763, Bernoulli 1771.

[364] Entry for Mar. 16, 1763 in Lalande 1763.

[365] Cassini to Ramsden, January 6, 1788; Ramsden to Cassini, Jan. 25, 1788; quoted in Wolf 1902: 289-292.

[366] Breteuil to Cassini, Apr. 24, 1788, quoted in Wolf 1902: 294.

[367] Cassini to Ramsden, Jan. 6, 1788, quoted in Wolf 1902: 289.

[368] Gillispie 1980: 127-130.

[369] Widmalm 1990.

[370] Roy 1790: 111-112.

[371] RSJB, Feb. 25, 1790.

[372] Ramsden to Banks, May 13, 1790, RSMM 3.30.

[373] Hornsby to unknown correspondent, no date, but after 1782, RASR A1.44.

[374] Hornsby to Ramsden, no date, but after Nov. 1785, RASR A1.70; Hornsby to the Duke of Marlborough, Oct. 31, 1794, RASR A2.134; Jan. 19, 1783, RASR A2.48. Ramsden's shop was in Piccadilly.

[375] Hornsby to the Duke of Marlborough, Apr. 22, 1784, RASR A2.77; Apr. 6, 1791, RASR A2.131(1); Apr. 11, 1795, RASR A2.139(2).

Bibliography of Primary Sources

Bacon 1960: *The new organon*, Francis Bacon, trans. & ed. Fulton H. Anderson, Macmillan, New York, 1960.

Bernoulli 1771: *Lettres astronomiques*, Jean Bernoulli, Paris, 1771.

Boscovich 1785: *Opera pertina ad opticam, et astronomiam*, Roger Boscovich, *Vol. 1*, Bassano, 1785.

Bougainville 1772: *A voyage round the world ... in the years 1766 ... [to] 1769*, Louis de Bougainville, trans. John [Johann] Forster, London, 1772.

Bradley 1728: "A letter to Dr. Edmond Halley giving an account of a new discovered motion of the fix'd stars," James Bradley, *Phil. Trans.*, 1728, *35*:637-661.

Bradley 1734: "Observations made in London by Mr. George Graham, F.R.S., and at Black River in Jamaica, by Colin Campbell Esq. F.R.S., concerning the going of a clock to determine the difference between the lengths of isochronal pendulums in those places," James Bradley, *Phil. Trans.*, 1734, *38*:302-314.

Bradley 1748: "A letter to the Earl of Macclesfield concerning an apparent motion observed in some of the fixed stars," James Bradley, *Phil. Trans.*, 1748, *45*:1-43.

Bradley 1832: *Miscellaneous works and correspondence of the Rev. James Bradley, D.D., F.R.S.*, ed. S. P. Rigaud, Oxford 1832.

Campbell 1747: *The London tradesman*, R. Campbell, London, 1747.

Cassini 1785: *Extrait des observations astronomiques et physiques à l'Observatoire Royal*, Jean Cassini, Paris, 1785.

Chambers 1728: *Cyclopaedia*, Ephraim Chambers, London, 1728.

Chambers 1786: *Cyclopaedia*, Ephraim Chambers, ed. Abraham Rees, London 1786.

Chambers 1820: *Cyclopaedia*, Ephraim Chambers, ed. Abraham Rees, London 1820.

Clairaut 1761: "Mémoire sur les moyens de perfectionner des lunettes d'approche, par l'usage d'objectifs composée de plusiers matières différemment réfringentes," Alexis-Claude Clairaut, *Mémoires,* 1756 (published in 1761), 380-437.

Cook 1777: *A voyage towards the South Pole, and around the world*, James Cook, London 1777.

Cook C1: *The charts and coastal views of Captain Cook's voyages, vol. 1: The voyage of the Endeavour*, ed. Andrew David, Hakluyt Society, London 1988.

Cook C2: *The charts and coastal views of Captain Cook's voyages, vol. 2: The voyage of the Resolution and the Adventure*, ed. Andrew David, Hakluyt Society, London 1992.

Cook J1: *The journals of Captain James Cook, vol. 1: The voyage of the Endeavour*, ed. J. C. Beaglehole, Hakluyt Society, Cambridge 1968.

Cook J2: *The journals of Captain James Cook, vol. 2: The voyage of the Resolution and the Adventure*, ed. J. C. Beaglehole, Hakluyt Society, Cambridge 1969.

Desaguliers 1734: *A course of experimental philosophy, vol. 1,* John Desaguliers, London 1734.

Dollond 1753: "A letter to James Short concerning a mistake in Mr. Euler's theorem for correcting the aberrations in the object-glasses of refracting telescopes," John Dollond, *Phil. Trans.*, 1753, *48*:289-291.

Dollond 1758: "An account of some experiments concerning the different refrangibility of light," *Phil. Trans.*, 1758, *50*:733-743.

Dollond 1789a: "Some account of the discovery which led to the grand improvement of refracting telescopes made by the late Mr. John Dollond F.R.S. in order to correct some misrepresentations in foreign publications of that discovery," Peter Dollond, 1789, RSL&P, IX.131

Dollond 1789b, "An answer to a paper presented to the Royal Society by Mr. Jesse Ramsden relating to the invention of the achromatic telescope," Peter Dollond, 1789, RSL&P, IX.146

Dollond 1789c: "Some account of the discovery made by the later Mr. John Dollond F.R.S. which led to the grand improvement of refracting telescopes .... With an attempt to account for the mistake in an experiment by Sir Isaac Newton," Peter Dollond, pamphlet, London, 1789.

Ellicott 1736: "The description and manner of using an instrument for measuring the degrees of expansion of metals by heat," John Ellicott, *Phil. Trans.*, 1736, *39*:297-299.

Ellicott 1752: "A description of two methods, by which the irregularity of the motion of a clock, arising from the influence of heat and cold upon the rod of a pendulum, may be prevented," John Ellicott, *Phil. Trans.*, 1752, *47*:479-494.

Euler 1747, "Sur la perfection des verres objectifs des lunettes," Leonhard Euler, *Opera omnia, 1911-1979. Vol. III.6 (Commentationes opticae, II)*, pp. 1-21, eds. Emil Cherbuliez and Andreas Speiser, Societatis Scientiarum Naturalium Helveticae, Basel, 1962.

Euler 1753: "Examen d'une controverse sur la loi de refraction de rayons de differentes couleurs par rapport à la diversité de mileux transparens par lesquels ils sont transmis," Leonhard Euler, *Opera omnia, 1911-1979. Vol. III.5 (Commentationes opticae, I)*, pp. 172-184, ed. David Speiser, Societatis Scientiarum Naturalium, Helveticae, Basel, 1962.

Euler 1765: "Remarques de M. Euler sur quelques passages qui se trouvent dans le troisième volume des opsucles mathématiques de M. d'Alembert," Leonhard Euler, *Opera omnia, 1911-1979. Vol. III.6 (Commentationes opticae, II)*, pp. 172-177, eds. Emil Cherbuliez and Andreas Speiser, Societatis Scientiarum Naturalium Helveticae, Basel, 1962

Euler 1768: *Lettres à une princesse d'Allemagne*, Leonhard Euler, *Opera omnia, 1911-1979. Vol. XII (Volumen posterius),* eds. Emil Cherbuliez and Andreas Speiser, Societatis Scientiarum Naturalium Helveticae, Basel, 1960

Euler 1823: *Letters to a German princess, with an introduction by Sir David Brewster*, Leonhard Euler, trans. Henry Hunter, Edinburgh, 1823.

Flamsteed 1835: *An account of the Rev'd John Flamsteed ... compiled from his own manuscripts*, ed. Francis Baily, London 1835.

Foster 1659: *Miscellanea... Mathematical lucubrations of Mr. Samuel Foster*, Samuel Foster, trans. John Twysden, London 1659.

Forster 1757: "On the number of people in England," Richard Forster, *Phil. Trans.*, 1757, *50*:457-465

Fuss 1783: "Éloge de Monsieur Léonard Euler," Nicholas Fuss, pamphlet, St. Pétersbourg, 1783.

Graham 1722: "Eclipse observed in Fleetstreet," George Graham, *Phil. Trans.*, 1722, *32*:198.

Graham 1724: "An account of observations made of the variation of the horizontal needle at London, in the latter part of the year 1722, and the beginning of 1723. By Mr. George Graham, Watchmaker, F.R.S.," George Graham, *Phil. Trans.*, 1724, *33*:96-107.

Graham 1726: "A contrivance to avoid the irregularities in a clock's motion occasion'd by the action of heat and cold on the rod of a pendulum," George Graham, *Phil. Trans.*, 1726, *34*:40-46.

Harris 1710: *Lexicon technicum*, John Harris, London, *Vol. 1*, 1704; *Vol. 2*, 1710.

Hawkesworth 1773, *An account of the voyages undertaken for making discoveries in the Southern Hemisphere*, John Hawkesworth, London 1773.

Herschel 1827: "Light", John Herschel, article in *Encyclopaedia metropolitana*, London, 1827.

Hooke 1665: *Micrographia: Or some physiological descriptions of minute bodies made by magnifying glasses*, Robert Hooke, London 1665.

Hooke 1674: "Animadversions on the first part of the machina coelestis of ... Hevelius ... together with an explication of some instruments ...," Robert Hooke, Cutlerian lectures, in *Early science in Oxford*, ed. R. T. Gunther, *Vol. 8*, Oxford University Press, Oxford, 1931.

Hooke 1935a: "The diary of Robert Hooke, 1672-1680", eds. H. W. Robinson and W. Adams, Taylor and Francis, London, 1935.

Hooke 1935b: "[Robert] Hooke's diary for late 1688 to March 1690, and December 1692 to August 1693," in *Early science in Oxford*, ed. R. T. Gunther, *Vol. 10*, Oxford University Press, Oxford 1935.

Hornsby 1775: *Lectures on experimental philosophy*, Thomas Hornsby, London 1775

Hume 1989: *Essays: Moral, political, and literary, 1741-1777*, ed. Eugene F. Miller, Liberty Classics, Indianapolis, 1989.

Kelly 1808: *The life of John Dollond, F.R.S., inventor of the achromatic telescope*, John Kelly, London 1808.

Lalande 1763: *Journal d'un voyage en Angleterre*, Jérôme Lalande, Paris 1763.

Lalande 1771: *Astronomie*, Jérôme Lalande, Paris 1771.

La Pérouse 1799: *A voyage round the world*, Jean-François de Galaup de la Pérouse, ed. M. L. A. Milet-Mureau, London 1799.

Ludlam 1786: "An introduction and notes on Mr. Bird's method of dividing astronomical instruments," William Ludlam, pamphlet, London 1786.

Martin 1759: *New elements of optics: Or the theory of the aberrations, dissipation, and colours of light*, Benjamin Martin, London 1759.

Maskelyne 1782: *Astronomical observations, made at the Royal Observatory at Greenwich, 1765-1774*, Nevil Maskelyne, *Vol. 1*, London 1782.

Maupertuis 1738a: *La figure de la terre, determinée par les observations de Messieurs de Maupertuis, Clairaut, Camus, Le Monnier, et de la M. L'Abbé Outhier, accompagnes de M. Celsius*, Pierre de Maupertuis, Paris, 1738.

Maupertuis 1738b: *The figure of the earth, determined from observations made by order of the French king at the polar circle*, Pierre de Maupertuis, London, 1738.

Montucla 1802: *Histoire des Mathématiques*, J. F. Montucla, *Vol. 3*, Paris, 1802

Newton 1999: *The principia: Mathematical principles of natural philosophy*, Isaac Newton, trans. I. B. Cohen and Anne Whitman, ed. I. B. Cohen, University of California Press, Berkeley, 1999.

Priestley 1772: *The history and present state of discoveries relating to vision, light, and colours*, Joseph Priestley, London, 1772.

Pringle 1773: "A discourse on the different kinds of air," John Pringle, pamphlet, London 1773.

Ramsden 1777: "Description of an engine for dividing mathematical instruments," Jesse Ramsden, pamphlet, London, 1777.

Ramsden 1779: "Description of an engine for dividing strait lines on mathematical instruments," Jesse Ramsden, pamphlet, London 1779.

Ramsden 1789: "Some observations on the invention of the achromatic telescopes," Jesse Ramsden, 1789, RSL&P, IX.138.

Ramsden 1790: untitled article, by "Veritas" [Jesse Ramsden], *Gentleman's magazine*, Sept. 30, 1790, *60*: 890-891.

Reynolds 1797: *The works of Sir Joshua Reynolds*, Joshua Reynolds, ed. Edmund Malone, London, 1797.

Roy 1790: "An account of the trigonometrical operation whereby the distance between the Royal Observatories of Greenwich and Paris have been determined," William Roy, *Phil. Trans.*, 1790, *80*:111-270.

Saint-Fond 1784: *A journey through England and Scotland to the Hebrides in 1784*, Faujas de Saint-Fond, London 1784.

Selva 1787: *Sei dialoghi ottici teorico-practica*, Lorenzo Selva, Venice, 1787.

Smeaton 1754: "Description of a new pyrometer," John Smeaton, *Phil. Trans.*, 1754, *48*:598-613.

Smith 1738: *A compleat system of optics*, Robert Smith, Cambridge, 1738.

Smith 1976: *An inquiry into the nature and causes of the wealth of nations*, Adam Smith, eds. R. H. Campbell and A. S. Skinner, Clarendon Press, Oxford, 1976.

Sprat 1667: *The history of the Royal Society of London, for the improving of natural knowledge*, Thomas Sprat, London, 1667.

Streete 1661: *Astronomia carolina: A new theory of the celestial motions*, Thomas Streete, London 1661.

Wilson 1799: *Reports of the Common Pleas and Kings Bench, 1742-1774*, G. Wilson, London 1799.

Bibliography of Secondary Sources

Babbage 1830: *Reflections on the decline of science in England*, Charles Babbage, London 1830.

Baker 2009: "The business of life: The socio-economics of the 'scientific' instrument trade in early modern London," Alexi Baker, in *Generations in towns: Succession and success in pre-industrial urban societies,* eds. F-E. Eliassen & K. Szende, Cambridge Scholars Publishing, Newcastle upon Tyne, 2009.

Baker 2010: "This ingenious business: the socio-economics of the scientific instrument trade in London, 1700-1750," Alexi Baker, DPhil dissertation, Oxford University, 2010.

Bazerman 1988: *Shaping written knowledge: The genre and activity of the experimental article in science*, Charles Bazerman, Wisconsin University Press, Madison, 1988.

Beaglehole 1968: "Introduction," J. C. Beaglehole, in Cook J1.

Beaglehole 1974: *The life of Captain James Cook*, Stanford University Press, Palo Alto, 1974.

Bektas & Crosland 1992: "The Copley Medal: The establishment of a reward system in the Royal Society," M. Yakup Bektas and M. Crosland, *Notes and records of the Royal Society of London,* 1992, *46*:43-76.

Bennett 1985a: "Instrument makers and the decline of science in England: The effect of institutional change on the elite makers of the early nineteenth century," J. A. [Jim] Bennett, in *Nineteenth-century scientific instruments and their makers*, ed. P. R. de Clerq, Museum Boerhaave, Amsterdam, 1985.

Bennett 1985b: "The scientific context," J. A. [Jim] Bennett, in *Nineteenth-century scientific instruments and their makers*, ed. P. R. de Clerq, Museum Boerhaave, Amsterdam, 1985.

Bennett 1986: "The mechanics' philosophy and the mechanical philosophy," J. A. [Jim] Bennett, *History of science*, 1986, 24: 1-28.

Bennett 1987: *The divided circle*, J. A. [Jim] Bennett, Sotheby's, 1987, London.

Bennett 1998: "Peter Dollond answers Jesse Ramsden," J. A. [Jim] Bennett, *Sphaera*, 1998, *8*:4-5.

Bennett 2011: "Mathematical Instruments," J. A. [Jim] Bennett, *Isis*, 2011, *102*:697-705.

Biagioli 1996: "Etiquette, interdependence, and sociability in seventeenth-century science," Mario Biagioli, *Critical inquiry*, 1996, *22*:193-238.

Bluhm 1958: "Remarks on the Royal Society's finances, 1660-1768," R. K. Bluhm, *Notes and records of the Royal Society of London*, 1958, *13*:82-103.

Boas Hall 1983: "Oldenburg, the *Philosophical Transactions*, and technology," Marie Boas Hall, in *The uses of science in the age of Newton*, ed. John G. Burke, University of California Press, Berkeley, 1983.

Boas Hall 1984: *All scientists now: The Royal Society in the nineteenth century*, Marie Boas Hall, Cambridge University Press, Cambridge, 1984.

Boas Hall 1991: *Promoting experimental learning: Experiment and the Royal Society, 1660-1727*, Marie Boas Hall, Cambridge University Press, Cambridge, 1991.

Boegehold 1943: "Zur vor- und frügeschichte der achromatischen fernrohrobjective," H. Boegehold, *Forschungen zur geschichte der optik*, 1943, *3*:81-114.

Brewer 1988: *The sinews of power: War, money, and the English state*, John Brewer, Knopf, New York, 1988.

Brown 1979: "Guild organization and the instrument-making trade, 1550-1830," Joyce Brown, *Annals of science*, 1979, *36*:1-34.

Bud & Warner 1998: *Instruments of science: An historical encyclopedia*, eds. Robert Bud and Deborah Warner, Garland, New York, 1998.

Burgess 1996: "The scandalous neglect of Harrison's regulator science," in *The quest for longitude*, ed. William J. H. Andrewes, Collection of Historical Instruments, Harvard University, Cambridge, Massachusetts, 1996.

Cardwell 1972: *The organization of science in England*, D. S. L. Cardwell, Heinemann, London, 1972.

Carter 1988: *The road to Botany Bay: An exploration of landscape and history*, Paul Carter, Knopf, New York, 1988.

Casanovas 1987: "Boscovich as an astronomer," Juan Casanovas, in *Bicentennial commemoration of R. G. Boscovich*, eds. M. Bossi and P. Tusi, Edizioni Unicopli, Milan 1987.

Chang 2004: *Inventing temperature: Measurements and scientific progress,* Hasok Chang, Oxford University Press, Oxford 2004.

Chapman 1990: *Dividing the circle: The development of critical angular measurement in astronomy*, Allan Chapman, Harwood, New York, 1990.

Christensen 1993: "Spying on scientific instruments: The career of Jesper Bidstrup," Dan Ch. Christensen, *Centaurus*, 1993, *36*:209-244.

Clark 1999: "The death of metaphysics in enlightened Prussia," William Clark, in *The sciences in Enlightened Europe*, eds. William Clark *et. al.*, Chicago University Press, Chicago 1999.

Clifton 1995: *Directory of British scientific instrument makers, 1550-1851*, Gloria Clifton, Philip Wilson, London, 1995.

Cochrane 1956: "Francis Bacon and the rise of the mechanical arts in eighteenth-century England," Raymond Cochrane, *Annals of science*, 1956, *12*:137-156.

Da Costa 2009: *The singular and the making of knowledge at the Royal Society of London in the eighteenth century,* Palmira Fontes da Costa, Cambridge Scholars Publishing, Newcastle upon Tyne, 2009.

Daston 1980: "Probabilistic expectation of rationality," Lorraine Daston, *Historia mathematica*, 1980, *7*:234-260.

Daumas 1953: *Les instruments scientifiques aux XVIIe et XVIIIe siècles*, Maurice Daumas, Presses Universitaires de France, Paris, 1953.

Daumas 1989: *Scientific instruments of the seventeenth and eighteenth centuries and their makers*, Maurice Daumas, trans. M. Holbrook, Portman Books, London 1989.

David 1988: "Introduction", Andrew David, in Cook C1.

Deiman 1994: "Optics and optical instruments, 1600-1800," J. C Deiman in *Companion encyclopedia of the history and philosophy of the mathematical sciences*, ed. I. Grattan-Guinness, Routledge, New York, 1994.

d'Espinasse 1956: *Robert Hooke*, Margaret d'Espinasse, William Heinemann, London, 1956.

Fara 1996: *Sympathetic attractions: Magnetic practices, beliefs, and symbolism in eighteenth-century England*, Patricia Fara, Princeton University Press, Princeton, 1996.

Feingold 2001: "Mathematicians and naturalists: Sir Isaac Newton and the Royal Society," Mordechai Feingold, in *Isaac Newton's natural philosophy*, eds. J. Z. Buchwald and I. B. Cohen, MIT Press, Cambridge, Massachusetts, 2001.

Fellman 1983: "Leonhard Eulers stellung in der geschicte der optik," Emil Fellman, in *Leonhard Euler, 1707-1783: beitrage zu leben und werk*, eds. Marcel Jenni and J. J. Burckhardt, Birkhauser, Basel, 1983.

Friedman 1992: *Kant and the exact sciences*, Michael Friedman, Harvard University Press, Cambridge, Massachusetts, 1992.

Galison 1997: *Image and logic: A material culture of microphysics*, Peter L. Galison, Chicago University Press, Chicago 1997.

Gillispie 1980: *Science and polity in France at the end of the old regime*, C. C. Gillispie, Princeton University Press, Princeton 1980.

Ginn 1991: "Philosophers and artisans: the relationship between men of science and instrument makers in London, 1820-1860,"

William Ginn, PhD dissertation, University of Kent at Canterbury, 1991.

Golinski 1992: *Science as public culture: Chemistry and enlightenment in Britain, 1760-1820*, Jan Golinski, Cambridge University Press, Cambridge, 1992.

Golinski 1999: "Barometers of change: meteorological instruments as machines of enlightenment," Jan Golinski, in *The sciences in Enlightened Europe*, eds. William Clark *et. al.*, Chicago University Press, Chicago 1999.

Good 1998: *Sciences of the earth: An encyclopedia of events, people, and phenomena*, ed. Gregory Good, Garland, New York, 1998.

Greenberg 1995: *The problem of the earth's shape from Newton to Clairaut: The rise of mathematical science in eighteenth-century Paris and the fall of 'normal' science*, John Greenberg, Cambridge University Press, Cambridge, 1995.

Hamilton 1926: *The English brass and copper industries to 1800*, Henry Hamilton, Longmans, London 1926.

Hankins 1985: *Science and enlightenment*, Cambridge University Press, Cambridge 1985.

Hankins & Silverman 1995: *Instruments and the imagination*, Thomas Hankins and Robert Silverman, Princeton University Press, Princeton, 1995.

Harris 1998: *Industrial espionage and technology transfer: Britain and France in the eighteenth century*, John Harris, Aldershot, Ashgate, 1998.

Heath 1981: *A history of Greek mathematics, vol. 1: From Thales to Euclid*, Thomas Heath, Dover, New York 1981.

Heilbron 1982: *Elements of early modern physics*, John Heilbron, California University Press, Berkeley, 1982.

Heilbron 1983: *Physics at the Royal Society during Newton's presidency*, John Heilbron, Williams Andrew Clark Memorial Library, Los Angeles, 1983.

Heilbron 1993: "A mathematicians' mutiny with morals," in *World changes: Thomas Kuhn and the nature of science*, ed. Paul Horwich, MIT Press, Cambridge, Massachusetts, 1993.

Hessenbruch 1999: "The spread of precision measurement in Scandinavia, 1660-1800," Arne Hessenbruch, in *The sciences in the European periphery during the Enlightenment*, ed. K. Gavroglu, Kluwer, Dordrecht, 1999.

Horrins 1830: *Memoirs of a trait in the character of George III*, Johann Horrins (pseud.), London 1830.

Howse 1975: *Greenwich Observatory, vol. 3: The buildings and instruments*, Derek Howse, Taylor and Francis, London 1975.

Hunter 1976: "The social basis and changing fortunes of an early scientific institution: An analysis of the membership of the Royal

Society, 1660-1685," Michael Hunter, *Notes and records of the Royal Society of London*, 1976, *31*:9-114

Hunter 1981: *Science and society in Restoration England*, Michael Hunter, Cambridge University Press, Cambridge, 1981.

Hutchison 1991: "Idiosyncrasy, achromatic lenses, and early romanticism," Keith Hutchison, *Centaurus*, 1991, *34*:125-171.

Iliffe 1993: "'Aplatisseur du monde et de Cassini': Maupertuis, precision measurement, and the shape of the earth in the 1730s," Rob Iliffe, *History of science*, 1993, *31*:335-375.

Jacob 1988: *The cultural meaning of the scientific revolution*, Margaret Jacob, Knopf, New York 1988.

Jacob 1997: *Scientific culture and the making of the industrial West*, Margaret Jacob, Oxford University Press, Oxford, 1997.

Jardine 1999: *Ingenious pursuits: Building the scientific revolution*, Lisa Jardine, Doubleday, New York, 1999.

Jungnickel & McCormmach 1999: *Cavendish: The experimental life*, Christa Jungnickel and Russell McCormmach, Bucknell University Press, Lewisburg, 1999.

Kiltz 1942: "Untersuchung zweier von Dollond und Ramsden hergestellter fernrohrobjective," G. Kiltz, *Zeitschrift für instrumentenkunde*, 1942, *62*:41-46.

King 1996: " 'John Harrison, clockmaker at Barrow; near Barton upon Humber; Lincolnshire': the wooden clocks, 1713-1730,"

Andrew L. King, in *The quest for longitude*, ed. William J. H. Andrewes, Collection of Historical Instruments, Harvard University, Cambridge, Massachusetts, 1996.

King 1955: *History of the telescope*, Henry King, Griffin, London, 1955.

Klein 1995: "Politeness for plebes: Consumption and social identity in early eighteenth-century England," Lawrence E. Klein, in *The consumption of culture, vol. 3: Image, object, text*, eds. John Brewer and Ann Bermingham, Routledge, London, 1995.

Kuhn 1976: "Mathematical versus experimental traditions in the development of physical science," Thomas Kuhn, in *The essential tension*, Thomas Kuhn, Chicago University Press, Chicago, 1976.

Langford 1989: *A polite and commercial people: England 1727-1783*, Paul Langford, Oxford University Press, Oxford 1989.

Lyons 1944: *The Royal Society: 1660-1940,* Henry Lyons, Cambridge University Press, London, 1944.

McKendrick 1982: "The consumer revolution in eighteenth-century England," Ian McKendrick, in *The birth of a consumer society: The commercialization of eighteenth-century England*, eds. Ian McKendrick *et. al.*, Indiana University Press, Bloomington, 1982.

Macleod 1987: "Accident or design? George Ravenscroft's patent and the invention of lead-glass," Christine Macleod, *Technology and culture*, 1987, *28*:776-803.

Macleod 1988: *Inventing the industrial revolution: The English patent system, 1660-1800*, Christine Macleod, Cambridge University Press, Cambridge, 1988.

McConnell 1992: *Instrument makers to the world: A history of Cooke, Troughton and Simms*, Anita McConnell, William Sessions, York, 1992.

McConnell 1994: "From craft workshop to big business: The London scientific instrument trade's response to increasing demand," Anita McConnell, *London journal*, 1994, *19*:36-53.

Mahoney 1996: "History of science," Michael Mahoney, in *The quest for longitude*, ed. William J. H. Andrewes, Collection of Historical Instruments, Harvard University, Cambridge, Massachusetts, 1996.

Mathias 1979: *The transformation of England: Essays in the economic and social history of England in the eighteenth century*, Peter Mathias, Columbia University Press, New York, 1979.

Millburn 1973: "Benjamin Martin and the Royal Society," John R. Millburn, *Notes and records of the Royal Society of London*, 1973, *28*:15-23.

Millburn 2000: *Adams of Fleet Street: Instrument makers to King George III*, John R. Millburn, Ashgate, Aldershot, 2000.

Money 1977: *Experience and identity: Birmingham and the West Midlands, 1760-1800*, John Money, Manchester University Press, Manchester 1977.

Morrell & Thackray 1981: *Gentlemen of science: Early years of the B.A.A.S.*, J. Morrell and A. Thackray, Oxford University Press, Oxford 1981.

Morrison-Low 2007: *Making scientific instruments in the Industrial Revolution*, A. D. Morrison-Low, Ashgate, Aldershot, 2007.

Morton & Wess 1993: *Public and private science: The King George III collection*, Alan Q. Morton and Jane A. Wess, Oxford University Press, Oxford 1993.

Musson & Robinson 1969: *Science and technology in the Industrial Revolution*, A. E. Musson and Eric Robinson, Manchester University Press, Manchester, 1969.

Nelthorp 1873: *A treatise on watch-work*, Henry Nelthorp, London 1873.

Nordenmark & Nordström 1938: "Om uppfinningen av den akromatiska och aplanatiska linsen," N. V. E. Nordenmark and J. Nordström, *Lychnos*, 1938, *4*:1-52.

Nordenmark & Nordström 1939: "Om uppfinningen av den akromatiska och aplanatiska linsen," N. V. E. Nordenmark and J. Nordström, *Lychnos*, 1939, *5*:313-384.

Pipping 1991: *The chamber of physics: Instruments in the history of sciences' collections of the Royal Swedish Academy of Sciences, Stockholm,* Gunnar Pipping, Stiftelsen Observatoriekullen, Stockholm, 1991.

Plumb 1982: "The acceptance of modernity," John Plumb, in *The birth of a consumer society: The commercialization of eighteenth-century England*, eds. Ian McKendrick *et. al.*, Indiana University Press, Bloomington, 1982.

Price 1980: "Philosophical mechanism and mechanical philosophy: some notes towards a philosophy of scientific instruments," Derek Price, *Annali dell'Istituto e Museo di Storia della Scienza di Firenze*, 1980, *5*:75-85.

Price 1984: "Of sealing wax and string," Derek Price, *Natural history*, 1984, *93*:49-56.

Pynchon 1997: *Mason and Dixon*, Thomas Pynchon, Henry Holt, New York, 1997.

Quill 1963: "John Harrison, Copley Medalist, and the £20,000 longitude prize," Humphry Quill, *Notes and records of the Royal Society of London*, 1963, *18*:146-160.

Robischon 1983: "Instrument makers in London during the seventeenth and eighteenth centuries," PhD dissertation, University of Michigan, 1983.

Rousseau & Haycock 1999: "Voices calling for reform: The Royal Society in the mid-eighteenth century - Martin Folkes, John Hill, and William Stukeley," G. S. Rousseau and David Haycock, *History of science*, 1999, *37*:377-406.

Rudwick 1985: *The great Devonian controversy: The shaping of scientific knowledge among gentlemanly specialists*, Martin Rudwick, Chicago University Press, Chicago, 1985.

Rusnock 1996: "Biographical introduction: James Jurin," Andrea Rusnock, in *The correspondence of James Jurin, 1684-1750, physician and secretary to the Royal Society*, ed. Andrea Rusnock, Rodopi, Amsterdam, 1996.

Salmond 1991: *Two worlds: First meetings between Maori and European, 1642-1772*, Anne Salmond, University of Hawai'i Press, Honolulu, 1991.

Schaffer 1983: "Natural philosophy and public spectacle," Simon Schaffer, *History of science*, 1983, *21*:1-43.

Schaffer 1988: "Astronomers mark time: Discipline and the personal equation," *Science in context*, 1988, *2*:115-145.

Schaffer 2011: "Easily cracked: Scientific instruments in states of disrepair," Simon Schaffer, *Isis*, 2011, *102*:706-717.

Shapin 1994: *A social history of truth: Civility and science in seventeenth-century England*, Steven Shapin, Chicago University Press, Chicago 1994.

Shapin & Schaffer 1985: *Leviathan and the air pump: Hobbes, Boyle, and the experimental life*, Steven Shapin and Simon Schaffer, Princeton University Press, Princeton 1985.

Shapiro 1979: "Newton's 'achromatic' dispersion law: Theoretical background and experimental evidence," *Archive for the history of the exact sciences*, 1979, *21*:91-128.

Sobel 1995: *Longitude: The true story of a lone genius who solved the greatest scientific problem of his age*, Dava Sobel, Walker, New York, 1995.

Solkin 1993: *Painting for money: The visual arts and the public sphere in eighteenth-century England*, David Solkin, Yale University Press, New Haven, 1993.

Sorrenson 1993: "Scientific instrument makers at the Royal Society of London," Richard Sorrenson, PhD dissertation, Princeton University, 1993.

Sorrenson 1995: "The state's demand for accurate astronomical and navigational instruments in eighteenth-century Britain", Richard Sorrenson, in *The consumption of culture, vol. 3: Image, object, text*, eds. John Brewer and Ann Bermingham, Routledge, London, 1995.

Sorrenson 1996a: "Towards a history of the Royal Society in the eighteenth century," Richard Sorrenson, *Notes and records of the Royal Society of London*, 1996, *50*:29-46.

Sorrenson 1996b: "The ship as a scientific instrument," Richard Sorrenson, in *Osiris*, vol. 11, 1996.

Sorrenson 1999: "George Graham, visible technician," Richard Sorrenson, *British journal for the history of science*, 1999, *32*:203-221.

Sorrenson 2001: "Dollond & Son's pursuit of achromaticity, 1758-1789," Richard Sorrenson, *History of science*, 2001, *34*:31-55.

Sorrenson & Burnett 1998: "Pyrometer," Richard Sorrenson and John Burnett, in *Instruments of science: An historical encyclopedia*, eds. Robert Bud and Deborah Warner, Garland, New York, 1998.

Steeds, 1969: *A history of machine tools*, William Steeds, Oxford University Press, Oxford, 1969.

Stewart 1992: *The rise of public science: Rhetoric, technology, and natural philosophy in Newtonian Britain*, Larry Stewart, Cambridge University Press, Cambridge, 1992.

Stewart & Weindling 1995: "Philosophical threads: Natural philosophy and public experiment among the weavers of Spitalfields," Larry Stewart and Paul Weindling, *British journal for the history of science*, 1995, *28*:37-62.

Stigler 1986: *The history of statistics: The measurement of uncertainty before 1900,* Stephen Stigler, MIT Press, Cambridge, Massachusetts, 1986.

Sutton 1995: *Science for a polite society: Gender, culture, and the demonstration of enlightenment*, Geoffrey Sutton, Westview Press, Boulder 1995.

Symonds 1969: *Thomas Tompion: His life and work*, R. W. Symonds, Spring Books, London 1969.

Taub 2009: "On scientific instruments," Liba Taub, *Studies in history and philosophy of science*, 2009, *40*:337-343.

Taub 2011: "Re-engaging with instruments," Liba Taub, *Isis*, 2011, *102*:689-696.

Taylor 1966: *The mathematical practitioners of Hanoverian England*, E. G. R. Taylor, Cambridge University Press, Cambridge, 1966.

Terrall 1992: "Representing the earth's shape: The polemics surrounding Maupertuis's expedition to Lapland," Mary Terrall, *Isis*, 1992, *83*:218-237.

Terrall 2002: *The man who flattened the earth: Maupertuis and the sciences of the Enlightenment*, Mary Terrall, Chicago University Press, Chicago 2002.

Thompson 1968: *The making of the English working class,* E. P. Thompson, Penguin, Harmondsworth, 1968.

Thomson 1812: *History of the Royal Society of London*, Thomas Thomson, London 1812.

Turner (Anthony) 1987: *Early scientific instruments: Europe 1400-1800*, Anthony Turner, Sotheby's, London, 1987.

Turner (Gerard) 1987: "Scientific toys," Gerard L'E. Turner, *British journal for the history of science*, 1987, *20*:377-398.

van Helden 1975: "The historical problem of the invention of the telescope," Albert van Helden, *History of science*, 1975, *13*:251-263.

van Helden 1983: "The birth of the modern scientific instrument, 1550-1700," Albert van Helden, in *The uses of science in the age of Newton*, ed. John G. Burke, University of California Press, Berkeley, 1983.

van Helden 1994: "Telescopes and authority from Galileo to Cassini," Albert van Helden, in *Osiris*, 1994, *Vol. 9*, 1994.

Wahrman 1995: *Imagining the middle class: The political representation of class in Britain*, c. 1780-1840, Dror Wahrman, Cambridge University Press, Cambridge, 1995.

Walters 1997: "Conversation pieces: Science and politeness in eighteenth-century England," Alice Walters, *History of science*, 1997, *35*: 121-154.

Warner 1990: "What is a scientific instrument: When did it become one, and why?" Deborah Warner, *British journal for the history of science*, 1990, *23*:83-93.

Weiner 1981: *English culture and the decline of the industrial spirit, 1850-1980*, Martin Weiner, Cambridge University Press, Cambridge, 1981.

Weld 1848, *A history of the Royal Society. Vol. 1*, London, 1848.

Weld 1858, *A history of the Royal Society. Vol. 2*, London, 1858.

Westfall 1980: *Never at rest: A biography of Isaac Newton*, Richard S. Westfall, Cambridge University Press, Cambridge, 1980.

Widmalm 1990: "Accuracy, rhetoric, and technology: The Paris-Greenwich triangulation, 1784-1788," Sven Widmalm, in *The quan-*

tifying spirit in the eighteenth century, eds. Tore Frängsmyr *et. al.*, University of California Press, Berkeley, 1990.

Willach 1996: "New light on the invention of the achromatic telescope objective," Rolf Willach, *Notes and records of the Royal Society of London*, 1996: *50*: 195-210.

Willach 1997: "The early history of the achromatic telescope objective," Rolf Willach, *Journal of the Antique Telescope Society*, 1997, *12*:4-13.

Wise 1993: "Mediations: Enlightenment balancing acts, or the technologies of rationalism," M. Norton Wise, in *World changes: Thomas Kuhn and the nature of science*, ed. Paul Horwich, MIT Press, Cambridge, Massachusetts, 1993.

Wolf 1902: *Histoire de L'Observatoire de Paris*, C. Wolf, Paris.

Woodbury 1972: *Studies in the history of machine tools*, Robert Woodbury, MIT Press, Cambridge, Massachusetts, 1972.

Woolf 1959: *The transits of Venus: A study of eighteenth-century science*, Harry Woolf, Princeton University Press, Princeton, 1959.

Index

Printed in Germany
by Amazon Distribution
GmbH, Leipzig